FUTURE**HYPE**

FUTURE**HYPE**

THE MYTHS OF TECHNOLOGY CHANGE

Bob Seidensticker

BERRETT-KOEHLER PUBLISHERS, INC.
San Francisco

Berrett-Koehler Publishers, Inc.
235 Montgomery Street, Suite 650
San Francisco, CA 94104-2916
Tel: (415) 288-0260   Fax: (415) 362-2512   www.bkconnection.com

ORDERING INFORMATION

Quantity sales. Special discounts are available on quantity purchases by corporations, associations, and others. For details, contact the "Special Sales Department" at the Berrett-Koehler address above.

Individual sales. Berrett-Koehler publications are available through most bookstores. They can also be ordered directly from Berrett-Koehler: Tel: (800) 929-2929; Fax: (802) 864-7626; www.bkconnection.com

Orders for college textbook/course adoption use. Please contact Berrett-Koehler:
Tel: (800) 929-2929; Fax: (802) 864-7626.

Orders by U.S. trade bookstores and wholesalers. Please contact Publishers Group West, 1700 Fourth Street, Berkeley, CA 94710. Tel: (510) 528-1444; Fax (510) 528-3444.

Berrett-Koehler and the BK logo are registered trademarks of Berrett-Koehler Publishers, Inc.

Printed in the United States of America

Berrett-Koehler books are printed on long-lasting acid-free paper. When it is available, we choose paper that has been manufactured by environmentally responsible processes. These may include using trees grown in sustainable forests, incorporating recycled paper, minimizing chlorine in bleaching, or recycling the energy produced at the paper mill.

Library of Congress Cataloging-in-Publication Data

Seidensticker, Robert B., 1957–
    Future hype : the myths of technology change / Bob Seidensticker.
        p.   cm.
    Includes bibliographical references and index.
    ISBN-10: 978-1-57675-370-0; ISBN-13: 978-1-57675-370-5
    1. Technology assessment.   2. Technology—Social aspects.   3. Technological innovations.
I. Title: Future hype, the myth of technology change.   II. Title.
T174.5.S45   2006
303.48'3—dc22   200505486

FIRST EDITION

11   10   09   08   07   06
10 9 8 7 6 5 4 3 2 1

Book producer: BookMatters, Berkeley; *Designer:* Bea Hartman; *Copyeditor:* Katherine Silver; *Proofreader:* Carrie Pickett; *Indexer:* Ken DellaPenta

To Bobby, Genny, and Sandy

Contents

Preface

We are as gods and might as well get good at it.
— STEWART BRAND, opening sentence
  of the *Whole Earth Catalog* (1968)

Most people feel certain that the pace of technological change increases exponentially. They think that the Internet and personal computers are only the most prominent of the many innovations that surge around us and that new ones arrive ever faster. They're certain that never before has the social impact of technological change been as profound or as pervasive as it is today.

But they are wrong.

The Internet isn't that big a deal. Neither is the PC. Abandon all technology and live in the woods for a week and see if it's your laptop you miss most. In fact, the technologies most important to us are the older ones—the car and telephone, electricity and concrete, textiles and agriculture, to name just a few. The popular perception of modern technology is inflated and out of step with reality. We overestimate the importance of new and exciting inventions, and we underestimate those we've grown up with. Change is not increasing exponentially. In fact, technology has disoriented and delighted for centuries. This book will attempt to recalibrate your thinking by looking at how technological change really happens.

Please don't misunderstand—I'm excited about the future possibilities of technology. And, of course, it *is* changing, and this change is often stressful: its impact and potential are so great that an accurate view is impressive enough—we needn't exaggerate. Let's overhaul our perception of technology change. We'll tear it down and build a stronger, more accurate model in its place.

This book is divided into two parts. In part I, I look at how and why we see technology incorrectly. I explore its downsides, how it bites back, its surprising fragility, and its unpredictability; I also review some tools and insights that will ease our sometimes tense relationship to it. I analyze and debunk nine "High-Tech Myths," fashionable but deceptive explanations for how technology works today. Once we begin to chisel away at the errors, a new and more accurate way of seeing technological change begins to emerge from the debris.

In part II, I look at the constancy of change in a broad range of areas—popular culture, health and safety, fear and anxiety, personal technologies, business; in all of these, history gives us repeated examples that make our experiences today seem unexceptional. This survey, illustrated with stories from thousands of years of human innovation, should lay to rest the notion that technology change is unique to our day. I draw most of my examples from the United States, not to ignore the importance of innovation in the rest of the world, but to focus the book. Nevertheless, the lessons here should be applicable to understanding technology change in other countries.

Just as a doctor who misdiagnoses a disease will provide the wrong treatment, our response to technology will be ineffective if we incorrectly perceive how it impacts society. Swept along by overexcitement with the new, we don't accurately see its promises or its weaknesses. My hope is that *Future Hype* will lead you to the clear vision needed to understand its true impact.

What could a clearer view provide? Knowing that technology

doesn't always deliver on promises, government and schools could be more rational and even skeptical before adopting it. Businesses might be sharper judges of technology and avoid the bandwagon effect. Worldwide, almost three *trillion* dollars are spent each year on information technology alone. A large fraction of that is wasted, but which fraction?

The view I offer is ultimately empowering—technology should answer to *us*. Readers who may not be encouraged by the cheery "and if you think it's changing fast now, just wait a few years!" will find here a breath of optimism. Learn how technology is *really* changing—and discover that it's much less scary than you've been told.

If people see technology more clearly, we would have a shrewder citizenry that would demand practical and constructive, rather than expedient or convenient, decisions from their politicians. They would be more able to analyze and discuss the relevant technology issues of the day—from the digital divide, to government support for space and other science programs, to national defense, to the value of computers in schools—and weigh more knowledgeably the pros and cons of what is being offered.

It's clear that many people care a lot about these issues. A recent National Science Foundation poll shows 92 percent of us moderately or very interested in new inventions and technologies. In one survey of the top news stories of the twentieth century—stories that included such fundamental events as the fall of the Berlin Wall, the start of World War II, and women's suffrage—fully 16 percent were about technology. Better-educated consumers would feel more confident about judging the value of a new product for themselves rather than relying on hype and would demand that it prove its value. They would know when the emperor had no clothes.

Over three decades ago, *Future Shock* by Alvin Toffler created a sensation by portraying technology spinning out of society's control. *Future*

Hype approaches the same topic but reaches a very different conclusion: that the popular view of technological change is wrong and the future won't be so shocking.

We live in a society exquisitely dependent on science and technology,
in which hardly anyone knows anything about science and technology.
—CARL SAGAN

Introduction:
Leveling the Exponential Curve

The further backward you look, the further forward you can see.
—WINSTON CHURCHILL

THE GAME OF CHESS DATES back to India fourteen hundred years ago. Legend says that the local ruler was so delighted by the game that he offered its inventor the reward of his choice. The inventor's request was defined by the game board itself: a single grain of rice for the first chess square, two for the next, four for the next, and so on, doubling with each square through all sixty-four. Unaccustomed to this kind of sequence, the ruler granted this seemingly trivial request. Little did he realize that the rice begins to be measured in cups by square fourteen, sacks by square twenty, and tons by square twenty-six. The total comes to about three hundred billion tons—more rice than has been harvested in the history of humanity.

Like the king in the chess story, most of us are inexperienced in this kind of exponential increase. Let's look at a present day example. In 1971, Intel introduced the 4004, its first microprocessor, with a performance of 0.06 MIPS (million instructions per second). Intel's Pentium Pro was introduced in 1995 with 300 MIPS, a five-thousand-fold performance increase in twenty-four years—about one doubling every two years. A car making the same speed increase would now have a top speed of about Mach 700. Give it another twenty-four years at the same rate of increase, and its top speed would exceed the speed of light.

Moore's Law, named after Intel cofounder Gordon Moore, predicts this exponential rise in computer performance: every two years, microprocessor speed doubles. Again. This law has been startlingly accurate for three decades, and the progress it predicts is expected to continue, at least for the near future. Because there is no precedent for this rapid performance improvement, we tend to view computers and their rapid change with wonder.

My own career of twenty-five years as a digital hardware designer and a programmer and software architect has been tied to Moore's Law. Ever since my high school years in the 1970s, I've been immersed in computer technology and have been an energetic cheerleader for technology in general. I was in awe of the change it brought about and was delighted to be a small part of that change. Change was exciting. And it was all around us—I grew up with the space program and jumbo jets, nuclear power and skyscrapers, *Future Shock* and *Mega-trends*. Exponential change seemed to be everywhere we looked.

To make sure we're all clear what exponential change looks like, figure 1 shows the differences between *no change, linear change*, and *exponential change*. The vertical axis is unlabeled—it could represent transistors in microprocessors, dollars for compound interest, the number of bacteria grown in a petri dish, or the grains of rice in the chess story. While they may start out slowly, exponential curves eventually snowball.

As I gained experience, I came to realize that change for its own sake wasn't as desirable for the software user as the software developer imagined. Users wanted new software to answer to bottom-line demands. Who would have guessed? Coolness alone was no longer enough—users demanded that software pull its weight, as they would for any other purchase.

They were right, of course. New software must provide sufficient additional benefits to outweigh the cost and aggravation of adopting it.

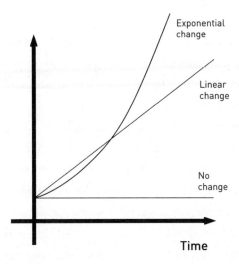

Figure 1. Exponential change contrasted with linear change and no change. The exponential curve doubles every time period. It might double every day if measuring bacteria growth or every decade if measuring number of miles of railroad track.

This is also true for other consumer products. The consumer might think: I like that digital camera, but it uses a new type of memory card. Will it become a standard or an unsupported dead end, like so many other products? Should I make MP3 copies of my favorite songs or keep them on CD? Is HDTV (High-Definition TV) really here, or is the current hype another false alarm? In general, is the latest hot product something that will last, or is it just a fad? The early adopters are quick to make this leap, but the chasm must be narrowed considerably for the majority of us. Change for its own sake wasn't as delightful as I'd thought, and I came to see things more from the user's perspective.

The high failure rate of new products challenges the inevitability of exponential change. A bigger challenge came as I studied high-tech products from the past, looking for precedents against which to compare my own projects. I wondered, why were these old products successful? and how could I apply what I learned to my own work? As I learned more about the history of technology, I was surprised to find examples that the exponential model could not explain. I gradually realized that there was a different way—a more accurate way—to look at such change.

The exponential model as a universal explanation for and predictor of technological change is at best an approximation and at worst a delusion. We can sustain it only by selecting just the right examples and ignoring all the rest. Technology does not always continuously improve. For example, commercial airplane speeds increased steadily for a while but halted when airlines realized that expensive super-sonic travel didn't make business sense. Highway speed limits increased steadily but also hit a ceiling. Record heights for skyscrapers increased rapidly during the first third of the twentieth century but have increased only moderately since then. Use of nuclear power has peaked, and manned space exploration halted after we reached the moon.

Specific areas of technology advance at different rates and come to the fore at different times. Cathedral building emerged during the 1200s while other technologies languished. Printing underwent dra-matic change in the late 1400s, then surged again in the early 1800s as mechanized presses provided cheap books and magazines. Steam power and mills had their heyday; later, it was electricity and electrical devices. There are dozens of examples of a specific technology surging forward and then maturing and fading back into the commonplace.

Perhaps the most venerable use of the exponential model has been to represent world population growth, but even here it's an imperfect metaphor. In the 1960s and '70s, experts warned that the world's pop-ulation was growing exponentially, and crowding would quickly get worse. Famine was just around the corner. Though dramatic, the model was inaccurate: world population growth is slowing and is expected to peak midcentury, and the populations of dozens of coun-tries are already *falling* in population (not counting immigration).

Despite the common perception, the impact of technology on society today is comparatively gentle. To see a truly serious example of the collision of technology and society, look at Britain during the Industrial Revolution almost two centuries ago. In 1811, armed gangs of Luddites smashed the textile machines that displaced their handmade crafts. Several years and over ten thousand men were required to put down the rebellion. The unrest spread to the Continent, where the word "sabotage" was coined—from the French word *sabot*, the wooden shoes used by workers to smash or jam machines. In the space of a generation, independent work on farms had given way to long six-day weeks in noisy and dangerous factories. Our own technological growing pains seem minor by comparison.

It's easy to focus on the recent at the expense of the old, but doing so can lead to a distorted view of our current situation. New products loom disproportionately large, often simply because they're new. The image of previous generations of Americans living quiet, static lives is fiction; they dealt with disruptions caused by technological innovations every bit as challenging and exciting as our own: the telegraph and electricity, the car and railroad, anesthesia and vaccines, concrete and steel, newspapers and mail. And if we go even further back, we see the fundamental developments on which society is based: agriculture, metallurgy, the beginnings of engineering, writing, textiles, transportation, timekeeping, basic tools and weapons, and so on. Are today's products really so amazing compared to those on which they were built? Too often we mistake a new technology for an important one.

Part of the problem is a narrow definition of *technology*. Obviously, the Internet, computer, and cell phone fit into this category. These are in the news and in our awareness. But this book will use a very broad definition of technology, including these new technologies as well as older and less glamorous ones mentioned above. Metallurgy, textiles, and all the rest were high tech at one point, they are still important to society, and examples from these older technologies will be liberally used in this book to illustrate that today's issues have, in fact, been around for a long time.

Sometimes the prevailing view of reality is an oversimplification. For example, small children are often taught that "All ocean creatures are fish." Though incomplete, it's a step in the right direction. When the children are a little older, we might teach them that all ocean creatures are fish—except whales and dolphins. When they are older still, we teach them that all ocean creatures are fish except marine mammals (like whales and dolphins), crustaceans (like crabs and lobsters), bivalves (like oysters and scallops), cephalopods (like nautilus and squid), and so on.

We frequently hear that the nature and rate of change in today's technologies are unprecedented. But like the fish simplification for children, this tells far less than the whole story; it helps explain some of what we see, but is inaccurate—and dangerously so. Leave behind the children's version of technology change, and explore how it is really affecting society and how it will impact us in the future.

> *We live in a technology-dense world. . . . We are terrifyingly naked*
> *without knowing elementary things about how [technologies] work.*
> —JOHN LIENHARD, *The Engines of Our Ingenuity* (2000)

PART I. THE WAYS WE SEE TECHNOLOGY INCORRECTLY

1 The Birthday-Present Syndrome

THE WRAPPING PAPER FLIES as Junior tears into his present from Grandma. It's the toy he's been hoping for, and he's delighted. All other possessions are forgotten as he begins to play with his new toy that will, in its turn, be ignored in favor of the next new thing.

When it comes to technology, most of us are like that kid with his birthday present—we are interested in the cool toy of the moment, and older technologies are only noticed in their absence. The result is that we don't see technology clearly; we don't soberly weigh today's new developments against the technologies we already have. The value of today's technology is inflated, and some revaluation is needed to restore a balance.

This chapter is an exercise in seeing more clearly the birthday-present syndrome, a seemingly permanent feature of our culture. It will also explore our uncomfortable coexistence with machines throughout the centuries. Society's relationship with technology is like a romance in which each person sees attractive traits in the other, but with familiarity comes some unpleasant surprises. Maybe she chews with her mouth open or has disagreeable political opinions. Maybe he's a slob or has antiquated views of a woman's role in society. Similarly, a technology is never pure and innocent, incorruptible in every one of its applications. We find bad traits along with the good; we adopt a technology hoping we will be pleased with the balance.

Good surprises can be difficult as well. We want to off-load tasks to machines, but egos can get bruised in the process. Does this new ability encroach on humanity? Are we reduced in value somehow by the success of our machines? Expect more of these kinds of questions as computers are increasingly able to do things that require thought; let us not forget, however, that this friction between society and technology has been around for a long time.

Technology Good and Bad

Humankind is either on its way to the stars
or hurtling out of a high-rise window to the street
and mumbling, "So far, so good."
—Edward Tenner,
   Why Things Bite Back (1996)

An ancient Chinese story tells of a farmer who owns a famous racehorse. One day, the horse runs away. His friends commiserate with him, but the farmer replies, "This isn't necessarily a bad thing." Soon, his horse returns and brings another fine-looking horse. His friends congratulate him, but the farmer observes, "This isn't necessarily a good thing." Later, the farmer's son is thrown while trying to tame the new horse. He breaks his leg, which leaves him lame. The farmer's friends offer condolences, but he responds, "This isn't necessarily a bad thing." Sure enough, war breaks out and the son's lameness prevents him from being conscripted. Though many neighbors' sons are killed in the fighting, the farmer's son is spared. Sometimes it's hard to tell what's a good thing and what's a bad thing.

But perhaps we can be certain in some cases. For example, we can all agree that the insecticide DDT is bad. The landmark book *Silent Spring*, by Rachel Carson (1962), made DDT's environmental crimes common knowledge. And yet DDT's discoverer won a Nobel Prize for his work in 1948, just six years after its properties were understood, and DDT was credited with saving five million lives by 1950. In the 1950s and '60s, DDT cut malaria in India to fifteen thousand cases per year, down from one hundred *million*. Given this remarkable progress, worldwide eradication of malaria seemed a strong possibility. Despite a growing understanding of the problems of resistance, environmental damage, and impact on human health, abandoning this insecticide was not the obvious course. Malaria kills millions of people per year even today, and DDT is still used in countries holding almost half of the world's population, including China, India, and Mexico. So, what's the moral? Is DDT a killer or a lifesaver? We could ask the same about antibiotics and vaccines—they mercifully saved lives and yet threatened widespread famine by encouraging dramatic overpopulation.

Kranzberg's First Law helps to clarify this situation: technology is neither good nor bad—nor is it neutral. At the risk of spoiling its Zen-like nature, let me propose an interpretation: a technology isn't inherently good or bad, but it *will* have an impact, which is why it's not neutral. Almost every applied technology has impact, and that impact will have a good side and a bad side. When you think of transportation technologies, for example, do you think of how they enable a delightful vacation or get the family back together during the holidays—or do you think of traffic jams and pollution? Are books a source of wisdom and spirituality or a way to distribute pornography and hate? Do you applaud medical technology for curing plagues or deplore transportation technology for spreading them? Does encrypted e-mail keep honest people safe from criminals or criminals safe from the police? Are plastics durable conveniences or everlasting pollutants? Counterfeiting comes with money, obscene phone calls come with the telephone, spam comes with e-mail, and pornography comes with the Internet. Every law creates an outlaw.

Opposites create each other. You can't have an up without a down, a magnetic North Pole without a South Pole, or a yin without its opposite yang. Providing a technology for a good use opens the door for the bad. Werner von Braun observed, "Science does not have a moral dimension. It is like a knife. If you give it to a surgeon or a murderer, each will use it differently." The same could be said for applications of technology.

The dilemma of finding and maximizing technology's gifts while minimizing its harm is especially important today, but it has plagued society for centuries. Today we worry about junk on the Internet; yesterday we worried about junk on TV (and before that, junk through radio and film and books and newspapers). Today we worry about terrorists using bioengineering techniques to make new diseases; yesterday we worried about the telegraph and railroad being used to conduct the Civil War. Today, computer pioneer Bill Joy has argued that because of the downsides of possible accidents, we should deliberately avoid certain areas of research; yesterday Leonardo da Vinci destroyed plans for devices like the submarine, anticipating their use as weapons.

Man Versus Machine Contests

Now the man that invented the steam drill
He thought he was mighty fine.
But John Henry drove fifteen feet
The steam drill only made nine.
—"John Henry" (folk song)

One particular kind of social friction caused by technology occurs when machines perform tasks that have traditionally been done by human beings. This is like a junior employee taking over the menial parts of your job—it's okay at first, but where will it end? Will it eventually cost you your job? Society has long been uneasy with machines encroaching on human turf, not just because of job loss, but also because of a vague loss of dignity. Could machines get uppity and forget their place?

The most direct example of this friction is the one-on-one turf battle—may the best man (or machine) win. Consider the story of John Henry. Though subsequently mythologized, he was a real person who worked on the Big Bend railroad tunnel in West Virginia in 1870. As a steel driver, he hammered long drills into the rock face to make holes for explosives. A mechanical drill had recently replaced steel drivers at other tunnels, and the drill manufacturer wanted it used on this project. Would it perform any better than men on the type of rock at Big Bend? To find out, a contest was proposed that pitted John Henry, the team's best driver, against the steam drill. John Henry defeated the steam drill but died in the process, thus celebrating the heroism of humanity while foreshadowing the ultimate futility of the man versus machine contest for physical tasks.

Perhaps the most prominent recent man versus machine contest was the defeat of chess grandmaster Gary Kasparov by IBM's Deep Blue computer. A computer as world chess champion had been "ten years away" since the 1950s, but not until 1997 did those ten years finally pass. After the Deep Blue victory, the press reported much soul-searching, as if humanity had been dealt a major blow. However, the fact that Deep Blue didn't celebrate its victory—and couldn't—underscores that it is a world-class chess player but nothing more. The original 1949 paper outlining the basics of computer chess noted that if human opponents didn't like how their game was progressing, they could always pull the plug.

To better understand the gulf that computers must still cross to be comparable to a human, imagine pitting a computer against a child rather than a chess champion. The computer's goal would be to match the child's understanding of the world. Some questions could test simple facts about the world (the sky is blue, water is wet, chairs are often made of wood), and others could examine common sense (What happens if you hit a pot with a spoon? What kinds of chairs burn? Can you stand on a table?). The ultimate test of this sort is the Turing Test, proposed by British mathematician Alan Turing in 1950, in which an observer communicates with two unseen entities, a computer and a human being. If the observer can't tell the difference, the computer has fooled the observer and passed the test. Present computer technology is a long way from passing this test, one far harder than a chess match.

Acting Like a Human

That this toil of pure intelligence . . .
can possibly be performed by an unconscious machine
is a proposition which is received with incredulity.
—COLUMBIA UNIVERSITY PRESIDENT,
   commenting on the adding machine (circa 1820)

Sometimes machines are deliberately designed to mimic how human beings work; a better approach is usually to discard those constraints and create a design that takes advantage of what machines do best. The history of printing gives us a good example. By the early 1800s, steam presses printing thousands of pages per hour were advancing the printing revolution Gutenberg began in 1455. The slow process of typesetting, however, remained a bottleneck. Even after text could be composed on a typewriter by the 1870s, each tiny metal character of type still had to be hand placed by skilled typesetters for printing. Not unlike programmers in the 1980s and '90s, fast typesetters could move between jobs at will and demand excellent wages. The best typesetters were celebrities and races became popular, attracting large audiences as if they were sporting events. Some competitors could set five thousand characters of justified and corrected text in an hour—better than one character per second. This was a tough job for machines to duplicate. Should they mimic the steps humans used or try a machine-specific approach?

By the 1880s, first generation mechanical typesetters were in use. Mark Twain was interested in this new technology and invested in the Paige typesetter, backing it against its primary competitor, the Mergenthaler Linotype machine. The Paige was faster and had more capabilities. However, the complicated machine contained eighteen thousand parts and weighed three tons, making it more expensive and less reliable. As the market battle wore on, Twain put more and more money into the project, but it eventually failed in 1894, largely because the machine deliberately mimicked how human typesetters worked instead of taking advantage of the unique ways machines can operate. For example, the Paige machine re-sorted the type from completed print jobs back into bins to be reused. This impressive ability made it compatible with the manual process but very complex as well. The Linotype neatly cut the Gordian knot by simply melting old type and recasting it. After investing a quarter of a million dollars in the project, Twain was bankrupt. He spent the next four years lecturing to repay his debts. (Twain's conclusion: never invest when you can't afford to and never invest when you *can*.)

As with typesetting machines, airplanes also flirted with animal inspiration in their early years. Flapping-wing airplane failures, however, soon yielded to propeller-driven successes. Airplanes don't fly like birds, and submarines don't swim like fish. Wagons roll rather than walk, and a recorded voice isn't replayed through an artificial mouth. A washing machine doesn't use a washboard, and a dishwasher moves the water and not the dishes. Asking whether a computer can think or wonder is like asking whether a car can trot or gallop—a computer has its own way of operating, which may be quite different from the human approach. The most efficient machines usually don't mimic how humans or animals work.

We can approach the question of thinking another way: Does a tree falling in a forest with no one to hear it make a sound? That depends on how *sound* is defined. Similarly, whether a computer duplicating a particular human skill is thinking or not depends on how *think* is defined. You could say that a computer chess champion doesn't think because it doesn't operate the way people do; or you could say that it thinks in its own way because it obviously gets the job done. To take another example, ELIZA was a famous 1965 computer program that played the role of a psychiatrist. It was so convincing that some users earnestly poured out their problems to the imagined intelligence, even though replicating ELIZA is simple enough to be assigned as homework in a college artificial intelligence course. Marvin Minsky considered artificial intelligence "making machines do things that would be considered intelligent if done by people."

Is the Turing Test still the ultimate test of cognition? Or is mimicking a human irrelevant as long as the computer gets the job done? In the movie *2001*, we see the computer HAL pass a second-generation Turing Test: not only is he convincingly human in conversation, he also becomes paranoid and homicidal. Perhaps acting like a human isn't such a worthy goal after all.

The gap separating computers and human beings is one of appearance as well as intelligence. The computer as an anthropomorphic robot that travels on two legs, manipulates things with fingers, and has the same approximate shape as a human has a long history, predating the 1950s low-budget sci-fi movies. The *Wizard of Oz* novel series introduced the robot Tik-tok around 1910, and an early robot appeared in the movie *Metropolis* (1927). The word *robot* was introduced into English from a Czech play in 1921. Fascination with smart machines extends back at least to the automaton orchestra built for a Chinese emperor over two thousand years ago.

One of the most famous historical automatons was actually a deception. The chess-playing "Turk" was unveiled in 1770. It toured Europe and defeated most opponents, including Benjamin Franklin. Charles Babbage's bout with the Turk stimulated his interest in computing machines. The Turk continued playing for decades, and few suspected its secret: a chess master hidden inside that controlled the turban-wearing mannequin. Elektro, "the amazing Westinghouse Moto-Man," was a seven-foot-tall robot exhibited at the 1939 New York World's Fair. Also a deception, a hidden operator controlled Elektro's speech. In a decision that seems especially dated now, its creators thought that the ability to smoke a cigarette added to its humanness.

Robots' real success so far has been in factories where precision and repeatability are important and appearance and adaptability are not. Machines work best when we let them be themselves. Around the house, the science fiction robot remains a dream, and yet telephone answering machines, microwave ovens, and other appliances have already encroached on the turf of the home robot.

The Ever-Moving Goal

"A slow sort of country!" said the Queen.
"Now, here, you see, it takes all the running you can do
to keep in the same place. If you want to get somewhere else,
you must run at least twice as fast as that!"
—Lewis Carroll, *Through the Looking-Glass* (1871)

Ask a magician to reveal how a trick is done. If you aren't told that it's a professional secret, you'll probably hear, "Actually, you really don't want to know." Knowing the secret eliminates the mystery and ruins the fun. Is fire-walking a mysterious example of mind over matter, or is it simple physics—that charcoal doesn't conduct heat well, so quickly moving feet don't get burned? (And which answer makes the more interesting story?) Similarly, the idea of a machine able to beat a chess grandmaster was magical and exciting, at least until it was achieved. Now we see it simply as an impressive feat but one without any impact on daily life. After all, as we now know, a dedicated chess computer can *only* play chess.

When you're told how a feat of illusion works, magic is replaced by mechanics and the fun is gone. When a computer reaches a human intelligence metric, it seems to show human-like qualities—that is, until you look behind the curtain and see very nonhuman algorithms and hardware.

A future technology milestone (the ability to see or to understand speech, for example) is sometimes considered proof of some aspect of humanity. But technology bears the burden that once that milestone is reached, it becomes a parlor trick. This new capability may well be useful, but it's no threat to humanity. An "electronic brain" from the 1940s performing thousands of additions per second certainly achieved a superhuman feat, yet a computer performing billions of additions per second today is not even noteworthy. Construction equipment that is as capable as hundreds of workers? Boring. Enormous factories that shape massive metal beams or make chemicals in ways humans could never duplicate? Ho-hum. Robotic assembly-line workers? Ancient history. Chess champion of the world? We thought that would be impressive, but have changed our minds—sorry. That which is "human" is redefined as machines approach it, like the mechanical rabbit that is always just out of reach of the racing greyhounds. For technology, the race is like the Red Queen said: "It takes all the running you can do to keep in the same place."

Perhaps that's the most important difference between man and machine. Society changes and improves, setting new goals once old ones are reached. But machines do what they're designed to do and no more. At least for now, it takes man to invent the next machine.

Technological Myopia: Revisiting the Birthday-Present Syndrome

Anything that was in the world when you were born
is normal and natural.
Anything invented between when you were 15 and 35
is new and revolutionary and exciting,
and you'll probably get a career in it.
Anything invented after you're 35
is against the natural order of things.
—DOUGLAS ADAMS

The world's first escalator was installed in Harrod's department store in London in 1889, and brandy and smelling salts were available to passengers made faint by the ordeal. It is hard for us to put ourselves in the places of people seeing for the first time, as adults, technologies that we have grown up with.

Try to remember the first time you used various technologies. For example, I remember the first time I flew on a Boeing 747, the first time I used a microwave oven, and the first time I used a mainframe computer. Other firsts for me: using an ATM to get cash in another state; participating in a videoconference call; and using a computer, a cell phone, and a Web browser. I remember the first time I saw a CD-ROM as the prize inside a cereal box.

By contrast, I do *not* recall the first time I rode in a car, watched television, read a book, used an electrical appliance, or made a telephone call. By the time I was born, these technologies had become unremarkable parts of society.

My kids will have a different list of unremarkable technologies. They have grown up with compact discs, personal computers, videotape, and cellular phones. For them, listening to music from a CD is commonplace but from a vinyl record is remarkable; I remember when it was the reverse. Similarly, flying in a jet plane for me is commonplace, but in a propeller-driven plane is noteworthy; my parents remember when it was the reverse. My grandparents knew a time when driving in a car was exciting, but horse-drawn transportation was not.

Joel Birnbaum observed: "Only people born before a technology becomes pervasive think of it as a technology; all others consider it part of the environment." This technological myopia—the tendency to see the new out of proportion to its impact and to discount the old— helps explain the pervasive and distorted view of technology in our society today. For a similar viewpoint, consider Saul Steinberg's well-known "A View of the World from Ninth Avenue." This *New Yorker* cover from 1976 shows several carefully drawn New York City streets in the foreground, with detail quickly dropping off in the distance. Beyond the Hudson River is a featureless and unimportant landscape composed of the rest of the United States, the Pacific Ocean, and Asia. In a similar way, we clearly see the changes caused by the PC, the Internet, and other recent technology, but older technologies, such as the printing press, train, and telegraph, fade into the distance. (By the way, I use

"PC" to refer to *any* personal computer, not just the IBM-compatible kind.)

For a different perspective, let's suppose we learned to communicate with dolphins. We could eventually ask, "So, what's it like to be wet all the time?" The dolphin might wonder what we are talking about. We understand *wet* because we understand *dry*. A dolphin wouldn't notice wetness even though it is constantly wet—in fact, *because* it is constantly wet. Similarly, we are so immersed in our technology that trying to evaluate today's society from the vantage point of today is inherently difficult, like any type of self-analysis, and it's not surprising that the common perception is off the mark.

We not only dismiss older technologies, we've also become accustomed to some rather startling consequences, things that might shock an outsider. For example, there are more than forty thousand car-related deaths in the United States annually. This is seen as an important but unremarkable fact of modern life. By contrast, when an airplane crashes and kills forty people, it becomes front-page news. This is the expected and accepted contrasted with the unexpected and surprising. Only the new is news.

In the Monty Python movie *Life of Brian*, there is a debate among the revolutionaries about the impact of Roman rule on Palestine. It sounds similar to our own debate about the relative importance of old and new technology. Here is a version of that technology debate, in *Life-of-Brian* style.

> BOSS: Technology today is so revolutionary! It makes what came before seem trivial. The Internet, the PC, cellular telephony—what technology from the past can hold a candle to this?
>
> LACKEY 1: Uh . . . the printing press?
>
> BOSS: Oh yeah. That is quite old, isn't it? Yes, that's certainly important.

LACKEY 2: And electricity.

BOSS: Yeah, OK. I'll grant you the printing press and electricity are two important old technologies.

LACKEY 3: And the telephone.

BOSS: Well, sure, obviously the telephone. I mean, that goes without saying, doesn't it? But apart from the printing press, electricity, and the telephone . . .

LACKEY 1: How about antibiotics and vaccines?

LACKEY 2: Agriculture and animal domestication.

LACKEY 1: Oh—railroads, cars, and airplanes.

LACKEY 3: And roads, dams, buildings, bridges—that sort of thing.

LACKEY 2: Uh—books, newspapers, mail delivery . . .

BOSS: All right, all right. But apart from the printing press, electricity, the telephone, and the foundations of medicine, agriculture, transportation, civil engineering, and communication, *what* has technology from the past ever done for us?

Anything that can be automatically done for you can be automatically done to you.
　　　　—DAVID WYLAND's Law of Automation

2 The Perils of Prediction

"DROP IN BY ROCKET PLANE ON TOTTENVILLE, the sootless garden city where you'll live in scientific comfort in AD 2000. You'll eat food from sawdust, shop by picture-phone, [and] cook on a solar range." This vision from 1950 also predicts vacuum-tube electronics, automation controlled by holes punched in a roll of paper, and houses built of plastic and metal. Family helicopters are common, dirty plates dissolve in hot water and are rinsed down the drain, and influenza and other ailments are no longer complaints.

Everyone wants to know what the future will be like, but as we can see, accuracy is not always possible. In this chapter, we'll explore the art of prediction and consider ways to see it more clearly. I can't give clairvoyance, but I do hope to point out some of the constraints on predictions and offer insights to evaluating them. Predicting is a very difficult undertaking: one thorough analysis of past predictions concluded that no more than a quarter of them were accurate. Or to be more specific: Are we trying to achieve more accurate predictions or be sharper in our assessment of those we hear, or both?

Poor Predictions

Prediction is very difficult, especially about the future.
—NIELS BOHR, physicist

Predictions are indeed difficult, but predictions about technology seem especially prone to error. Here are some famous failures.

There is no reason for any individuals to have a computer in their home. (Ken Olson, founder of Digital Equipment Corp., 1977)

Television won't matter in your lifetime or mine. (*Radio Times* editor Rex Lambert, 1936)

The radio craze will die out in time. (Thomas Edison, 1922)

This "telephone" has too many shortcomings to be seriously considered as a means of communications. The device is inherently of no value to us. (Western Union internal memo, 1876)

Rail travel at high speed is not possible, because passengers, unable to breathe, would die of asphyxia. (Dr. Dionysus Lardner, professor at University College London, 1823)

What can we conclude from this list? Obviously even the experts are too timid when predicting how technologies advance. Take courage, use your imagination, and see a bold new future!

But there is another category of wrong predictions, a larger and more influential category: the *over*predictions. These are the dangerous predictions, the ones that stick in our minds and support the myth of exponential change.

- Marie Curie predicted that radiation would prolong life (this was in 1904). Ironically, she died from leukemia due to overexposure to radiation.

- All trees in the United States will be gone by 1920, cut down for heating and cooking (1890).

- Fast electric ships will cross the Atlantic in two days (1900).

- Atomic energy would "transform a desert continent, thaw the frozen poles, and make the whole world one smiling Garden of Eden" (1908).

- Thomas Edison predicted, "In 15 years, more electricity will be sold for electric vehicles than for light" (1910).

- Animal parts (a chicken breast, for example) will be grown separately, without the need to raise the whole animal (1932).

- Buckminster Fuller imagined cities housed under domes (1965).

- We'll have moon bases and passenger rockets to the moon by 1980 and robot soldiers by 1990 (1966).

- Electromagnetic fields are so beneficial that classrooms will be deliberately enveloped in them to help students remember better (1980s).

Let's try a thought experiment to see how hard it is to make a successful long-range prediction. Ben Franklin, who lived from 1706 to 1790 and was a man of quick and inquisitive mind, once wrote that he wished to wake up in the future. Suppose we could give him his wish. Before we watch him as he marvels at the twenty-first century, however, let's take advantage of his naive view of our world. We'll give him a list of impressive developments since his day, some of which have actually come about, and some that haven't. Would he be able to tell them apart?

The list could include instant worldwide voice communication, electricity to power household lighting and appliances, and flying machines that travel five hundred miles per hour. From the not-here-yet category, we could mix in mental telepathy, the ability to speak with the dead, and houses built of materials other than wood. We could include anesthetics and organ transplants, the ability to non-invasively see inside the body, and medicines that prevent or cure the worst diseases he knows, such as plague, yellow fever, and smallpox; add to that a one-day cure for broken bones, a medicine that removes fat, and cures for arthritis and the common cold. We could then throw in the moon landing, the hydrogen bomb, and advance warning for natural disasters mixed with underground cities, the extermination of mosquitoes and similar pests, and humans bred for specific characteristics just like crops and domestic animals.

We quickly see some as old news and the rest as speculation, but could Franklin do the same? I don't think so. Why can we prevent smallpox but not colds? Why do we build one-hundred-story skyscrapers out of glass and steel but houses out of wood? Who would have guessed that in the twenty-first century, JFK airport outside New York still can't find a better way to keep birds from runways than falcons?

If we were in Franklin's position and given a list of future predictions, how would *we* tell the winners from the losers? For example, Microsoft was just another small company with a big dream until the PC became hot. If you'd seen their business plan among a thousand others in 1975, would you have singled them out for greatness?

One way to explain the poor record of predictions is with Amara's Law, offered by Roy Amara of the Institute for the Future, which states that we overestimate short-term changes and underestimate long-term changes. The short-term part is pretty easy to explain: when a new technology comes along and begins to catch on, it gets a lot of press. Much of the talk is necessarily speculation or hype, establishing expectations the technology can't possibly meet.

Underestimating the long-term changes means underestimating how thoroughly today's technology will eventually insinuate itself into our lives. Looking backward, electricity, cars, the telephone, and other mature technologies are everywhere today, in more places than could have been predicted. It may be more relevant, though, to say that by the time we reach that "long-term" point somewhere in the future, the impact from the technology in question *will not be noticed.* For example, the pervasiveness of electricity and other mature technologies today would be impressive only to people at the dawn of those technologies—now they're taken for granted, and *we* don't care.

Another aspect to long-term change, as the Franklin example illustrates, is that when a technology is completely new (in the lab or before) and not an extrapolation of a product that exists in the market today, long-term changes that result from the technology are almost impossible to predict. In 1880 predictions about how flight would affect us by 1920 were not off only by a matter of degree (they missed the number of airplanes by a factor of ten, say), they were completely off target. Similarly, today's impressive new developments such as the Internet and the PC weren't *under*estimated forty years ago—they weren't estimated at all. They weren't even on the soothsayers' radar.

Here's a summary of this updated law assessing how we will predict, or "mis-predict," technological change. Imagine the year is 2010 and we're making predictions for 2015 (short term) and 2040 (long term).

1. In 2010, we will overestimate the impact that our new technology will have by 2015.

2. In 2040, we will find that we underestimated how pervasive 2010 technology would become, but no one will care: what was new and exciting in 2010 is ignored in 2040, because something else has become the exciting new technology du jour.

3. In 2040, we will see that our 2010 attempts at long-term predictions of technology *not yet present* in 2010 will be completely wrong.

A thorough discussion of how to make good predictions and spot the bad ones would take a complete book. But because so much of technology hype comes from predictions, some guidelines for evaluating them follow. For more on this subject, I suggest *Megamistakes* by Steven Schnaars.

Don't Get Stuck in the Present

One of the problems with predictions of the future
is that by the time it's clear that they have had little resemblance
to actual events, it's too late to get your money back.
—RAY KURZWEIL,
  The Age of Spiritual Machines (1999)

The first step in evaluating predictions is to discount those that assume exponential change. If exponential change were widespread, we would find that most predictions underestimate, and reality would outpace the prediction. And yet the opposite is true. What's changing exponentially in many cases is expectation, not technology.

Predictions are often more a picture of the present rather than the future. The 1960s and '70s saw predictions of nuclear-powered planes and vacations in space because nuclear power and space exploration were the hot topics. The workweek was shrinking, so predictions about work in the future were also common. Predictions about the population explosion dominated other dire scenarios, but they were wrong because they assumed that the issues, priorities, and concerns of the present would continue unchanged into the future.

We heard about depletion of energy reserves and environmental degradation. Clearly, society had to use less energy. During the oil embargo of 1973–74, an assertion that anxiety about energy use was just a passing fancy would have been seen as ill-informed and even irresponsible—but would have been correct. Once the pressure was off, the issue faded from view. At this writing (2005), oil prices have shot up and we care again.

In 1980 *Newsweek* magazine predicted that robots could replace at least half of U.S. factory workers within twenty years. Of course, many factory workers *were* replaced, but most of the lost jobs were outsourced to other workers, not robots. A 1966 forecast envisioned robot tractors, indoor farms, irrigation with desalinized seawater, and synthetic meats. It missed the real farming issues such as fluctuating prices and a move away from beef for health reasons. In 1967 the future of merchant shipping looked nuclear. Actually, the real issues were increasing international competition and container ships. An 1893 forecast projected a massive expansion of the railroad as well as pneumatic tubes to carry both mail and people, completely missing the huge impact of the car and the airplane.

Thoughtless extrapolations are another danger of an exaggerated fixation on today's issues. If today there are jet planes, tomorrow there must be supersonic planes, and a plane's capacity will increase to a thousand passengers. If today has television, tomorrow must have 3D or holographic TV, and telephones will become videophones. If today many diseases are under control, tomorrow we will be able to control them all, and life spans will reach one hundred years. If today we have modoet weather prediction, tomorrow we will predict weather a year ahead. And if we've just finished the Empire State Building, mile-high buildings are next.

Mark Twain had some comments about careless extrapolation. During his time, a number of engineering projects straightened and shortened the winding Mississippi River. He speculated about this trend.

In the space of 176 years the Lower Mississippi has shortened itself 242 miles. This is an average of a trifle over one mile and a third per year. Therefore, any calm person, who is not blind or idiotic, can see that in the Old Oolitic Silurian Period, just a million years ago next November, the Lower Mississippi River was upward of 1,300,000 miles long, and stuck out over the Gulf of Mexico like a fishing-rod. And by the same token any person can see that 742 years from now the Lower Mississippi will be only a mile and three-quarters long, and Cairo [Illinois] and New Orleans will have joined their streets together, and be plodding comfortably along under a single mayor and a mutual board of aldermen. There is something fascinating about science. One gets such wholesale returns of conjecture out of such a trifling investment of fact.

Tomorrow will look more like today than most predictions would lead us to believe.

Avoid Technology Infatuation

No sensible decision can be made without taking into account
not only the world as it is, but the world as it will be.
—Isaac Asimov, author

Technology is often thrilling, but too often we let our excitement cloud our projections. People still use products for basic reasons, and a new product must work with those reasons. When evaluating a prediction, challenge all assumptions, avoid over-complex logic, and use common sense.

Polaroid instant photography was pretty amazing. So were Polavision instant home movies. But Polavision was a short-lived experiment and Polaroid photography was never more than a niche. As amazing as technology can be, that alone doesn't make for a successful product.

One 1967 view of the future home saw disposable dishes and inflatable furniture. Don't worry about spills—just hose down your plastic furniture and let the water run through a drain in the floor. But just because a product is possible, doesn't mean people will want to buy it.

Scott Paper sold paper dresses beginning in 1966, and five hundred thousand were shipped in six months. You could shorten your dress with scissors, customize it with paint, and discard it after wearing it once. It was a fad whose popularity lasted only a few years, like CB radios a decade later. Neither answered real customer demands.

Predictions about the success of these products were based on such infatuation with the technology itself that they didn't consider whether the consumer would actually care. Other proposals that never inspired enough demand were underwater hotels, artificial moons for lighting cities at night, dehydrated or irradiated food, moving sidewalks, geodesic domes, cars that drove themselves, and paperless offices.

Even big companies make mistakes. GE and Motorola were invested in CB radio, AT&T and Sears in videotex (a precursor to the Internet), and the *New York Times* and RCA in fax newspapers (a mini newspaper sent by radio and printed in each home). In the final analysis, consumers just didn't need what these products provided.

The fax newspaper provides a good case study of the results of infatuation with technology. For a few years in the late 1930s and in another burst of enthusiasm ten years later, dozens of radio stations broadcast them. Some were published four or more times daily to give readers the latest news. David Sarnoff of RCA saw this to be as promising a new technology as television, and one journalism school offered a class in fax newspaper production. Newspapers' attitudes toward the product were identical to those of many companies toward the early Internet: we're not sure how to make money in this business, but if we don't jump in, we might miss something big.

Too often, forecasts about a fledgling industry begin with the assumption that success is inevitable. The only question remaining is: What growth curve best documents that success? And yet, success is not inevitable; it's not even likely. Most new products fail.

A good forecast shows analogies to past successful products, but it also examines failed products to show why the new product won't be like them (see figure 2). Successful products follow an S-curve: an S-shaped graph of product sales over time that shows slow growth initially, fast growth as the product becomes mainstream, and slowing growth as the market becomes saturated. A forecast made in advance of sales, however, and assuming that sales will follow an S-curve is very optimistic since an S-curve only applies to *successful* products. The assumption of fast growth for the videophone or supersonic passenger airplane or any other failed product gave the wrong answer no matter how clever the presentation or how powerful the supercomputer that helped with the analysis.

A long and complex argument with fancy analysis and statistics is another warning sign of erroneous predictions: this usually obscures more than it illuminates. The fundamental market factors are still pretty simple. Who are the potential customers, and how many are there? What product or process will be replaced and why is the new approach better? Where is the new approach worse (for example, do you require customers to change their habits)? What social trends work to the new product's favor or detriment (changing concern for health or finances, for example)? Does the benefit outweigh the cost? Don't forget to ask if the forecast came from a possibly biased source like a company or industry group that would benefit if the prediction came

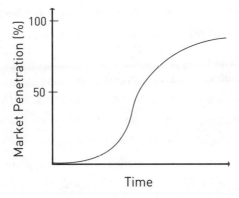

Figure 2. S-curve. Note the rapid growth in the middle.

true. There are more questions like this, of course, but common sense takes you a long way to avoiding technology infatuation.

New Products Don't Win on Every Point

"But he has nothing on!" at last cried all the people.
The Emperor writhed, for he knew it was true.
—HANS CHRISTIAN ANDERSEN,
   "The Emperor's New Clothes" (1837)

In the short story "The Man Who Came Early," by Poul Anderson, a U.S. Army soldier stationed in Reykjavík is mysteriously transported one thousand years back in time.

Struggling in this strange land to find something he is skilled at and with which he can repay his new friends, he describes to the Vikings new boats using triangular sails that allow them to sail into the wind. Surely this is better than the Viking longboat whose square sails are effective only when the wind is at its back and must be rowed when moving in any other direction. But the Vikings point out a number of deficiencies. The new boat's deep keel would prevent it from going up a shallow river or being beached, often done to find shelter from storms and protection against attack. In addition, there were no docks or piers to provide access to such sea-bound boats. The weighted keel would also be too difficult to make. In short, the marvelous new invention had far too many drawbacks.

A new technology is rarely superior to an old one in every feature. It can improve with time, of course, but out of the gate it's not the obvious winner. We find many examples of this with modern products. MP3 songs are convenient, but the sound quality is worse than compact discs (it's worse even than that of a new vinyl record). Laptops must be plugged in or frequently recharged, and they're heavy and expensive, unlike paper and pencil. Computer monitors have just 5 percent of the resolution (dots per inch) of a high-quality book or magazine. Computer LCD displays cost more and have a narrower viewing angle than monitors (CRTs). Digital video (DVDs and digital TV) has new visual anomalies not present with analog video (such as cable or VCR). Web pages aren't as high resolution as those in a paper catalog, and a Web site doesn't allow the equivalent of quickly leafing through a catalog. A David Sipress cartoon illustrates this retrenchment with "The Off-Line Store." Signs in the window read, "All items are actual size!" and "Take it home as soon as you pay for it!"

Another example of a new technology that is not necessary better is 3D movies. A screenwriter developing a script for such a movie must contrive scenes to show off 3D's benefits. I remember one 1981 movie with a scene showing nuts poured down a well. The camera shot was from below, and the audience saw the nuts whizzing past them. This was definitely cool, but it had absolutely nothing to do with the story. The downsides to 3D films may be modest (not every theater is able to show them and viewers must wear special glasses), but the benefits still must be big enough to compensate. In its half century of existence, 3D hasn't provided them.

This is also the marketing challenge of the videophone. It's definitely cool, but it just isn't that useful. Proponents imagine Grandma marveling at her new grandchild from across the country or a similar situation where video adds a lot to a telephone call. But these instances are rare. The minor benefits don't outweigh the hassles of installing and using it. Perhaps only by making the videophone a prominent computer feature (and free) is it likely to get used.

Of course, MP3 players, calculators, laptops, digital video, and other products have all been successful. The point is that new technologies, even the successful ones, are superior to existing technologies on some features but worse on others. And the more features that compare poorly, the less likely a product will succeed. If you're surprised at the slow acceptance of a new product, make an objective comparison of how the product competes on *every* point to find the logic behind the market's indifference.

Consumers are surprisingly logical when evaluating products. Maybe a new audio player is both cheaper and higher quality than current players. Sounds like a winner, right? But consumers will want to know whether it supports their existing music library of CDs or MP3 files. If not, how much hassle and expense is it to convert to the new format? Are there players for the bookshelf, for the car, and for jogging? Can the new format be played on a PC? Will all music companies provide music in the new format? Must existing players be discarded and replaced with this new one? Consumers need compelling reasons to make such a switch.

Some industry watchers claim that this process is not always logical. They give as an example the Betamax video format. It had better video quality, and yet VHS became the dominant format. Did the market choose an inferior product in this case? No—Beta actually did *not* have noticeably better quality (*Consumer Reports* at the time showed it a toss-up) and, more importantly, VHS beat Beta on recording time. Another example: Did the market act illogically when it kept the ubiquitous QWERTY keyboard layout rather than the more logical Dvorak? No—QWERTY was entrenched when Dvorak came along, and Dvorak didn't offer a big enough improvement to outweigh the hassle of the change. If the new product is better (in a comparison of *all* relevant features), it'll sell. If it doesn't sell, it's not better. We must never forget to look at the big picture when predicting how a new product will do in the market. As Russell Ackoff said in his book *The Art of Problem Solving* (1978), "Irrationality is usually in the mind of the beholder, not in the mind of the beheld."

Finding the Next Big Thing

Asking the right questions is superior
to finding elaborate answers to the wrong questions.
—STEVEN SCHNAARS, *Megamistakes* (1989)

Where will the Next Big Thing come from? It's rarely from the company about to get hammered by it. It's hard to predict the future when we don't even know where to look.

The digital watch didn't come from the established watch companies. The calculator didn't come from the slide rule or adding machine companies. Video games didn't come from Parker Brothers or Mattel. Semiconductors didn't come from the vacuum-tube makers. The ballpoint pen didn't come from the fountain-pen industry. The Internet browser didn't come from Microsoft. Looking further back, cars didn't come from wagon makers, refrigerators from ice companies, or light bulbs from candle makers.

More recently, traditional phone companies had to buy their way into the cellular telephone business. The top players in paper directories, like Thomas Register and Yellow Pages, are *not* the top players in Web search engines. The leader in robot vacuum cleaners is iRobot, not Electrolux or Hoover.

This shouldn't be too surprising. The trick with a robot vacuum cleaner is the robot, not the vacuum cleaner. Skill in dreaming up board games doesn't carry over into skill in designing video games. Vacuum-tube makers were not particularly well placed to see the need for (or path to) the integrated circuit. The market leaders are probably where evolutionary products will come from, but this is less likely for more revolutionary products. And when it comes to a completely new kind of product—something that gets the job done in a completely new way—it usually comes from outside the industry. The legendary garage shop as a source of innovative products makes the job of prediction much tougher.

There seems to be a rule of constant value with predictions: longer-term predictions have more potential value and yet are less likely to come true; short-term predictions are more certain, but they don't tell us much new information. Long-term predictions can be a good starting point for a debate. We can then take steps to steer toward or away from that vision of the future as appropriate. But be skeptical in proportion to how far in the future the prediction tries to reach. Keep in mind that bubbly predictions form much of the foundation of today's hype.

When you get the urge to predict the future,
better lie down until the feeling goes away.
—*Forbes* magazine (July 10, 1978)

3 The Unintended Wager

JOHNNY APPLESEED WAS A REAL PERSON. Born John Chapman, he lived on the American frontier in Ohio, Indiana, and Illinois in the early 1800s. Unlike the legend, he didn't cast his apple seeds randomly around the countryside but rather planted orchards with the goal of selling the seedlings. He did, however, widely spread the seeds of fennel, not native to America, as well as those of other plants considered medicinal at the time. He thought that fennel cured malaria. It doesn't. In fact, it's now seen as an invasive pest.

We often don't get the results we expect. Sometimes the surprise is a good one and we get positive synergy. Sometimes copper plus tin gives bronze. Sometimes one plus one equals three. But too often, the technology we cast through society turns out like Johnny Appleseed's fennel and exhibits some nasty traits.

Adopting a new technology becomes an unintended wager. These unexpected consequences, both good and bad, are an important side effect of systems problems, and fitting technology into society is always a systems problem. A computer is a system of interacting parts, and a school is a system in which the computer is a part. A car is a system, as is the network of streets on which it drives. Because they are so complicated, systems can be tricky. Too often, applying a technology fix turns out to be like fighting a fire with gasoline. This chapter will conclude with some ways to see systems problems more clearly.

Unexpected Consequences in Health

Science invents conveniences by design
and inconveniences by accident.
—G. K. CHESTERTON

Humanity's impact on nature is often heavy-handed and clumsy and leads to unexpected and undesired consequences. One example is what happened when the World Health Organization sprayed DDT to kill mosquitoes and combat malaria in Borneo in the 1950s. The mosquitoes died and the malaria incidence dropped. Success!

But the story doesn't end there. This environmental model—that mosquitoes carry disease and DDT kills mosquitoes—was just a small part of a much larger ecosystem. Caterpillars lived in and fed on the thatched roofs of the local houses, but they were kept in check by parasitic wasps. When the DDT killed the wasps, the population of roof-eating caterpillars exploded. The DDT also killed flies and cockroaches, a good thing until the geckos that had kept the insect population in check ate the poisoned insects. Dying geckos could no longer escape the cats, and the cats in turn were poisoned and died. Without cats, villagers saw an enormous increase in rats, which ate grain and spread disease. What seemed a simple system was actually quite complex.

We can find many other such examples in modern medicine. Modern surgical techniques and pharmacology—vaccines, antibiotics, anesthesia—have prevented or cured countless illnesses as well as relieved symptoms and improved life. Ignoring their economic costs, advances in medical science seem to be only for the good, but have, in fact, led to some surprising negative consequences. For instance, the occasional summer epidemics of polio in the early 1900s were actually enabled by cleanliness. When infants of dirtier times were exposed to the ubiquitous poliovirus, the immunity they still had from their mothers helped them safely develop resistance. When society became cleaner and the virus rarer, people were more likely to be exposed for the first time as older children or adults, after they had lost their residual maternal defenses. As a result, the innocuous virus became a menace. Similarly, rubella was never a serious adult disease until it became rare in children. The "hygiene hypothesis" extends this thinking by speculating that the recent rise in asthma and allergy rates is also a result of cleanliness. The hypothesis states that juvenile immune systems need occasional exposure to a naturally dirty environment to avoid overreacting later to benign annoyances like mold or pollen.

The mother of all unintended consequences, however, may be the dramatically improved life expectancy brought about by modern medicine and clean living conditions, which in turn contributed to the shocking population explosion after the Industrial Revolution. World population was one billion by about 1800. It reached two billion by 1930 and three billion by 1960. It has more than doubled since then. Agricultural innovations came to the rescue with improved food yields, but we can easily forget how alarming this rapid increase was at the time. Paul Ehrlich in his book *The Population Bomb* (1968) expressed a widely held view: "In the 1970s the world will undergo famines— hundreds of millions of people are going to starve to death."

Technological developments can have an impact on health in other ways. The Black Death spread slowly over Europe, taking several years to move from the Black Sea west and north through Europe (and killing perhaps a third of the continent's population by 1353); today we have airplanes that can take a contagious person anywhere in a single day. Though safeguards are in place, we live with the risk that Ebola or SARS or a new strain of influenza could appear in our backyard, brought there by our own transportation system. Technology can also assist the spread of disease indirectly, for example, by providing mosquitoes with good breeding conditions in the still water behind dams and in discarded tires. Drug resistance in bacteria (like tuberculosis) or parasites (like malaria) makes some of our most impressive medical gains only temporary.

Hospitals themselves harbor surprising and unexpected dangers. Every year, two million patients get an infection from their stay in a U.S. hospital and one hundred thousand of them die. These infections have become the fourth leading cause of death. As Samuel Goldwyn reportedly said, "A hospital is no place to be sick."

Unexpected Consequences of the Internet

In adversity, everything that surrounds you is a kind of medicine
that helps you refine your conduct,
yet you are unaware of it.
In pleasant situations, you are faced with weapons
that will tear you apart,
yet you do not realize it.
—HUANCHU DAOREN, Taoist philosopher (circa 1600)

One of the greatest uses of the World Wide Web is as a universal medium of communication. Getting a message out through the Web is cheaper than through a television station, can reach a wider audience than a local newspaper, and is faster than printing a book. "The Cluetrain Manifesto" (1999) raved: "We embrace the Web not knowing what it is, but hoping that it will burn the org chart—if not the organization—down to the ground. Released from the gray-flannel handcuffs, we say anything, curse like sailors, rhyme like bad poets, flame against our own values, just for the pure delight of having a voice. And when the thrill of hearing ourselves speak again wears off, we will begin to build a new world. That is what the Web is for." A little overheated, perhaps, but this remains a popular view of the invigorating freedom the Web provides.

While the lack of constraint may indeed thrill the provider of Web content, it is a huge downside for the reader. The Web is perhaps the least reliable source of information available. With no editor or standard for most Web content, the reader is forced to burrow through this information landfill and separate the useful and accurate content from the misinformation, ads, porn, and irrelevant drivel. Information that can find no outlet other than the Web is third tier, and the writer's enthusiasm for seeing it in print is rarely shared by the reader. There are reliable sources, of course, but their reputations were usually made in the print or broadcast worlds (consider the *Encyclopaedia Britannica*, the *New York Times*, NBC News, and so on). The paradox is that the Web's publishing strength is also its weakness.

With e-mail we find a similar paradox. It's easy to pass along useful e-mail but just as easy to pass along junk. This has given us inboxes full of spam, not-very-relevant business correspondence sent FYI, and jokes and other nonsense. For example, e-mail chain letters are common. You've probably heard about the Neiman Marcus cookie recipe ("they overcharged me for this recipe so I'm taking revenge by making it public—pass it on"), an appeal with an impossible claim ("Bill Gates will send three cents to this charity every time you forward this e-mail—pass it on"), or a warning against some computer virus ("this is important news—pass it on"). Whether it's greed, an emotional appeal, or a chance to be a good citizen, these chain letters attempt to infect you with a motivation to send them along. There's probably no real computer virus behind that warning e-mail—the e-mail itself is the virus, and *you* are the intended host.

Even when the facts are accurate, an e-mail appeal can be impossible to rein in. In 1989 a boy with cancer had a dying wish to set the Guinness record for receiving the most get-well cards. He broke the record, he's quite well now, and after fifteen years and 350 *million* cards he'd really, really like people to stop sending them.

Through the Internet we can choose to read news from targeted news sources, avoiding stories we don't care about, but do we lose something important? In the same way that avoiding exposure to germs can lead to a weaker immune system, avoiding exposure to a random array of news can leave us poorly informed; we miss the opportunity to scan unfiltered headlines to find the small fraction of serendipitous and useful stories we would have missed otherwise.

The Web was invented to allow easy international collaboration among scientists, and yet here we find a similar problem. The bandwagon effect is stronger when ideas can be debated and a consensus reached earlier in the scientific process. Fringe ideas are dropped more quickly when they can be aired more easily. This may often be more efficient, but it can prevent surprising discoveries that the old-fashioned approach might have fostered.

And Unexpected Consequences Everywhere Else

The chief cause of problems is solutions.
—Eric Sevareid

Examples of the unexpected consequences of innovation abound in the world of business. The film industry fought VCRs, thinking that they would destroy its livelihood, but the movie industry's revenue continues to rise, and home video rentals and sales account for half of Hollywood's income. The music industry feared pirated music downloads, but CD sales weren't hurt as predicted. Banks resisted pro-consumer credit card laws, but those laws created the confidence that built the huge credit card industry. High tariffs coddle rather than strengthen domestic businesses (why lower prices when foreign competition is constrained?), and lax piracy laws discourage domestic innovation (why spend to innovate if competitors can steal your work?).

The imminent paperless office has been a frequent prediction since the 1970s, yet paper use has risen rather than fallen. American businesses now process one trillion pieces of paper annually. One Xerox executive has a new prediction: "We'll have paperless offices about the same time as we have paperless bathrooms." Similarly, telecommuting hasn't emptied office buildings or rush hour. True, many tasks previously carried out on paper are now done electronically and certain work that required an employee to be in an office can now be done from home or from the road. Nevertheless, the reality is often quite different from the prediction or, more surprisingly, the exact opposite.

The problem of imperfect technology becomes life threatening when that technology is civil engineering. The heroic Dutch boy plugging the leaky dike with his finger knew how dependent his country was on technology. Hundreds of square miles of the Netherlands have been reclaimed from the sea and are below sea level, and much of the rest is barely above sea level—all protected by dikes.

We find a recent example of technology dependence along the Mississippi River. Structures had been built to contain the river within its banks during frequent floods. Houses, businesses, and cities sprang up on the secured land behind the levees and flood walls. While this land was safe during most years, by encouraging development on historical floodplains, engineers had unintentionally increased the size of the problem when the defenses were eventually overwhelmed, as happened in the flood of 1993. St. Louis was particularly hard hit, with losses exceeding $10 billion. An even greater disaster was that caused by Hurricane Katrina, which hit New Orleans in August 2005. Much of the city is below sea level, and when levees failed, extensive flooding helped create the most expensive natural disaster in U.S. history.

Note that this is not about natural disasters (like the earthquake and tsunami that destroyed Lisbon in 1755) or failure-of-technology disasters (like the dam collapse that flooded Johnstown in 1889). This is when technology said, "Come on in, the water's fine!" and we did—and some of us drowned. As engineers try to insulate society from natural forces by preventing erosion in front of beachfront houses, providing water to desert communities, extinguishing natural fires that threaten homes, or making floodplains safer, they increase complacency and actually enable that rare but huge disaster.

Technology has had unexpected effects in the home as well. The washing machine's efficiency resulted in cleaner clothes, not fewer work hours. In fact, the amount of laundry done per week went up tenfold from the 1950s to the 1980s. Similarly, the vacuum cleaner got the house tidier but didn't free up cleaning time. Even when we include all the additional household conveniences introduced in the mid-

1900s—gas or electric stove (instead of wood or coal), clothes dryer, dishwasher, refrigerator, indoor plumbing, and so on—housework hours didn't drop in response.

We can find more such paradoxes if we look for them. Consumer electronics devices use more energy off than on—that little bit of power they need to respond to the remote control's power-on command, for example, adds up if you consider the much longer time they spend off. Does the Internet isolate people by eliminating the need to interact, or does it enable new connections based on shared interests rather than shared geography? Recycling was expected to profit from the goldmine of garbage that society produces, and yet dumping often costs less. Plastic's properties of strength and inertness are desirable when you're using it, but troublesome when you throw it away and want it to decompose.

The examples multiply if we reach a little further back in time. Lead improves gasoline's performance in cars and was introduced in the 1920s as a way to stretch scarce oil. Fifty years later, we were shocked to discover a close correlation between the amount of leaded fuel consumed and the amount of lead in the blood of American children.

CFCs (chlorofluorocarbons, or Freon) also have a long history. This group of chemicals is an important category of refrigerants and was introduced with a flourish at a 1930 conference. Inventor Thomas Midgley gently exhaled a lungful of Freon to extinguish a candle, demonstrating in a single breath that Freon was both nontoxic and nonflammable. The industry had been searching for safe alternatives to toxic or flammable refrigerants like ammonia or methyl chloride. Unfortunately, their very inertness means that CFCs stay in the atmosphere for a long time, where they attack the ozone layer.

In the early twentieth century, movies like *Modern Times* and *Metropolis* imagined a dehumanized future with people subservient to machines. But a few decades later, we saw work hours drop and the workplace become more civilized, largely thanks to machines. Social critics then warned of an excess of free time. What would we do with it all? *Future Shock* in 1970 stated that, "The work week has been cut by 50 percent since the turn of the century. It is not out of the way to predict that it will be slashed in half again by 2000." Today, with work hours going back up, we debate both extremes: on one hand that longer work hours are a fact of life, and on the other that technology will soon replace jobs, creating unemployment rather than free time.

As cars became increasingly popular in the early 1900s, it didn't take a genius to see the increasing need for infrastructure such as roads, gas stations, and repair shops. Billboards next to highways, traffic jams, and smog, however, would have been tougher to predict. More difficult still would have been to foresee developments like fast-food restaurants, drive-in movie theaters, malls displacing local shops, and the flight to the suburbs. Unfortunately, just because the indirect consequences of technology are sometimes more difficult to anticipate, doesn't make them any gentler on society. The unintended aspects of a technology product are features, just as the intended ones are.

Dealing with Systems

Perfection means not perfect actions in a perfect world,
but appropriate actions in an imperfect one.
—R. H. BLYTH, student of Zen

As these many examples show, components of a system—whether an environment, a machine, or a society—can interact in unexpected ways to produce surprising consequences. Systems are usually more complex than they initially seem, and we often aren't aware when we're even dealing with one. How, then, can we make sense of and deal with these systems? Lewis Thomas in his book *The Medusa and the Snail* offers this paradoxical warning about meddling with any system: "Whatever you propose to do, based on common sense, will almost inevitably make matters worse rather than better." He concludes that "the safest course seems to be to stand by and wring hands, but not to touch." Of course, when we already *have* meddled in a system—by introducing carbon dioxide or CFCs into the atmosphere or releasing nonnative species into an environment, for example—we may need (or at least be tempted) to meddle further to try to fix the damage. But, returning to Thomas's counsel, many new problems are caused by clumsily applied "solutions."

John Gall in *Systemantics* offers this warning about systems:

> Systems are seductive. They promise to do a hard job faster, better, and more easily than you could do it by yourself. But if you set up a system, you are likely to find your time and effort now being consumed in the care and feeding of the system itself. New problems are created by its very presence. Once set up, it won't go away; it grows and encroaches. It begins to do strange and wonderful things and breaks down in ways you never thought possible. It kicks back, gets in the way, and opposes its own proper function. Your own perspective becomes distorted by being in the system. You become anxious and push on it to make it work. Eventually you come to believe that the misbegotten product it so grudgingly delivers is what you really wanted all the time. At that point, encroachment has become compete. You have become absorbed. You are now a Systems-person.

Seeing systems for what they are is an essential first step. In the late 1960s, Horst Rittel distinguished between "tame" and "wicked" problems. This is not the difference between easy and hard problems— many tame problems are very hard, and wicked problems, while not evil, are tricky and malicious in ways that tame problems are not. The unexpected consequences we've discussed are systems problems, and they are wicked. We will understand systems better—and why they spawn such consequences—if we understand a little more about the properties of wicked problems and learn to approach them with the appropriate respect.

Tame problems can be clearly stated, have a well-defined goal, and once solved, stay solved. They work in a Newtonian, clockwork way. The games of chess and Go present tame problems. Wicked problems have complex cause-and-effect relationships, include human interaction, and imply inherently incomplete information. They require compromises. For example, mass transit is a wicked problem. Everyone likes mass transit—unless it comes through their neighborhood, consumes road lanes, or they have to pay for it. When there is a big difference between how something works in the lab, in academia, on paper, or in one's head and how it works in the real world and affects real people, you know you are dealing with a wicked problem.

Tame and wicked problems differ in many ways. See if the traits of wicked and tame problems as described below sound familiar, either with the examples mentioned here or with situations you have experienced yourself.

- *Problem Definition.* A tame problem can be clearly, unambiguously, and completely stated. Math problems are tame. By contrast, there is no absolute statement of a wicked problem. To state a wicked problem means to also state its solution. That is, the problem can't be stated without a proposed solution in mind, and coming up with a new solution means seeing the problem in a new way. Avoid locking in a problem definition too soon.

- *Goal.* A tame problem has a well-defined goal, such as the QED in a proof or the checkmate in chess. With a wicked problem, you could keep iterating and refining your solution forever—or go back and consider other solutions. After all, if a wicked problem is something you can't define, how can you tell when it's resolved? You don't stop because you're done (you've reached the goal) but rather because of external constraints (you've run out of money, time, or patience, for example). You must strive for an adequate solution, not a perfect one.

- *Evaluating Solutions.* Solutions are unambiguously correct or incorrect with tame problems. The solution to a wicked problem is not judged as correct or incorrect but somewhere in the range between good and bad.

- *Time.* The solution to a tame problem can be judged immediately (that is, there is no maturation time), and the problem remains solved. Euclid's geometry proofs are still valid today. Evaluating the solution to a wicked problem takes time (because the results of implementing the solution take time to be appreciated) and is subjective. Is that a good design? Maybe, but maybe not. Like the response to art, different people will have different answers, and the solution causes many side effects (unintended consequences). Additionally, a "solved" wicked problem may not stay solved. In fact, wicked problems are never really solved, they are merely addressed; they're treated, not cured. Your perception of the effectiveness of the solution may change over time.

- *Consequences.* Trial and error may be an inefficient approach to a tame problem, but it won't cause any damage. Implementing or publicizing a proposed solution doesn't change the nature of the problem. With a wicked problem, however, every implementation changes reality—it's no longer the same problem after an attempted solution, and the solution you realize you should have tried may now not work.

- *Reapplying Past Solutions.* A class of tame problems can often be solved with a single principle. A general rule for finding a square root or applying the quadratic formula will work in all applicable cases, for example. By contrast, the solution to a wicked problem is unique. We can learn from past successes, but an old solution applied unchanged to a new problem won't produce the old result; many unexpected consequences arise when we rush to reapply (without customization) a particular solution we've seen before.

- *Problem Hierarchy.* A tame problem stands alone; it is never a symptom of a larger problem. A wicked problem always is. For example, if the cost of something is too high, this can be a symptom of a higher-level problem, such as the company not having enough money. Often, we can't see the higher-level problem:

"This new software is terrific! I can't imagine what could be better."

Systems are large and complex and exist in the real world. Industry's dreams and expectations for its new high-tech products are formed in the lab, but it is in the system of society where they're put to use. None of this is to say that we can't address systems problems but that we should do so with caution and respect.

The 1954 short story "Answer" by Fredric Brown illustrates a final example of unexpected consequences. In it, Brown envisions many great scientists working for many years to build a giant computer network by connecting the computing power of billions of planets. As the inaugural question for this technological marvel, the gathered dignitaries ask, "Is there a God?"

The computer doesn't hesitate before answering, "There is now!"

Everything has both intended and unintended consequences.
The intended consequences may or may not happen;
the unintended consequences always *do.*
—DEE HOCK, president of VISA (1994)

4 If It Ain't Broke, Be Grateful

AN EXTENSIVE PROGRAM OF COPYING WORKS from around the known world endowed the celebrated library at Alexandria, Egypt, with a collection of about half a million manuscripts. When a Muslim army took the city in 640 CE, a caliph reasoned that any document that agreed with the Koran was redundant, and any that contradicted it was blasphemous. He ordered the library destroyed.

Historians wince at the thought of the priceless manuscripts lost to us as a result, but we have our own version of this story. Digital information is slipping through our fingers—not quickly in an inferno, but gradually and relentlessly all around us. CDs, disks, and tapes all have a surprisingly short lifetime. In theory, digital is forever, but in practice, our records are more short-lived than they've ever been.

In the last chapter, we saw how technology can surprise us with unexpected consequences. Now we turn to its errors and failures. Data is ephemeral, products are buggy, and networks are vulnerable to joyriding hackers or enemies of the state. Much is made of the two steps forward, but little of the one step back. Today's technology is impressive, but we must see it accurately, flaws and all.

Fragile Digital Storage

[The life of a skyscraper is] 50 years or so,
and that assumes assiduous maintenance.
—Vincent Scully,
 New York Times Magazine (1999)

In 1086, twenty years after William of Normandy conquered England in the Battle of Hastings, he commissioned a survey of his new dominion. This survey is now known as the Doomsday Book. In 1986, on the Doomsday Book's nine-hundredth anniversary, the BBC unveiled a £2.5 million updated version. With digitized photos, maps, video, and text—in all, contributions from about a million people—it was expected to stand next to its parchment predecessor as a fundamental piece of scholarship.

And yet the multimedia version is now unusable. Only a few of the custom PCs developed for the project still exist, and its twelve-inch videodiscs are unreadable on any other device. A research project was necessary to salvage the data and store it in a more accessible format. Problem solved? Hardly. No digital format will be readable forever, and preserving this data will be an ongoing task of copying and reformatting to adapt to changing technology. That's a lot of fuss when the original Doomsday Book, almost a millennium old, sits well preserved and available to researchers in a Public Record Office in London.

We turn from this virtual time capsule to a traditional one to see another example of the fragility of high technology. The *New York Times* completed its end-of-the-millennium time capsule in 1999. Because the creators wanted their "Times Capsule" to remain sealed for a thousand years, the artifacts' stability was especially important. A complete copy of the year's *New York Times* would be an obvious addition. It would be an afternoon's work to copy that onto a handful of CD-ROMs and toss them in with the rest of the artifacts. Barring any sort of catastrophe, civilization will continue to advance, and reading our laughably primitive CDs should be child's play for the people of the year 3000, right?

Not quite. The advanced people of one thousand years hence wouldn't even get the chance to interpret them because CDs only have a lifetime of a few decades; after a thousand years, they might as well have been put in there blank. Paper, even newsprint, is readable for much longer than a disc. The *New York Times Magazine* commented on this problem of digital evaporation: "Almost everything today gets recorded, yet almost nothing will survive."

You may have heard that "digital is forever," and it does have some excellent qualities. To see its benefits, imagine a long-distance telephone call sent over a wire, first in analog form and then as a digital equivalent of ones and zeroes. The analog transmission fades after a certain distance and must be amplified; the longer the distance, the more amplifications are required. Unfortunately, noise such as static accumulates with the signal, and both are amplified. The result is a call that sounds noisy roughly in proportion to the distance it traveled.

Digital transmissions also need amplification, but as opposed to analog, noise can be detected and eliminated. Every one or zero is turned from a noisy bit, haggard and worn from its journey, into a good-as-new bit, identical to the original. This noise-free data stream is then amplified and sent on its way. As a result, the quality of the digital call can be perfect regardless of the distance of the trip. For similar reasons, stored digital information (on a CD or in an MP3 file, for example) can be retrieved noise-free, which isn't possible with analog storage (on a record or tape, for example).

Digital storage has some tremendous advantages over its analog equivalent. In practice, however, tapes, disks, CDs, and other media are impermanent repositories for digital data. The inevitable degradation of CDs as the surface oxidizes has been called "CD rot." The magnetic bits on tape and disk also gradually fail as their tiny magnetic fields fade. There is another danger. Magnets cover refrigerators and the occasional filing cabinet, they're often unexpected features on the backs of kitchen timers and other gadgets, and they're even in telephone speakers. A magnet needs only a moment near a floppy disk or credit card magnetic strip to cause damage.

Cuneiform tablets and chiseled obelisks are clumsy vehicles for information storage, but they'll be legible long after our magnetic or optical media have turned into gibberish. For example, the 1960 U.S. census was recorded on magnetic tape. By 1975, none of that information could be read. NASA has tapes from past missions that it might not be able to copy before they deteriorate into illegibility. Compare this with text on acid-free paper that is readable for centuries (acid has been an unwanted element of paper since cheap papermaking was introduced in the late 1800s).

Even if the media—floppy disk, tape, and so on—are in good condition, they require a machine to read the data. Unfortunately, hundreds of models of disk readers and tape drives are now obsolete. Most of these were proprietary products that only occupied niches in the market, but major players such as the 8-inch and 5.25-inch floppy disks have also been superseded. This issue touched me personally when I was working on an out-of-date PC in the mid-1990s. I realized that its two different floppy drives (a 5.25-inch and a 3.5-inch) gave me what would probably be the last chance I'd get to retrieve files I had stored on 5.25-inch floppies. It was—I haven't seen a PC configured like this since. The chances of the data on a 5.25 floppy being readable and of finding a drive to read it are both small. At the time of this writing, 3.5-inch floppy drives are also becoming rare. The lesson here is that files must not only be copied to refresh the data, but the storage media may need to be updated to keep abreast of the latest technology.

A third problem with accessing old digital records is the impermanent nature of the format in which the data is stored. A data format might be proprietary like Microsoft Word's document file (.doc) or an open standard like the JPEG format (.jpg) used for images. Like a particular disk or tape drive model, a format will be in vogue for a while, perhaps even as the market leader, but will eventually be forgotten. WordStar was *the* word processing program in the early 1980s, but you'll be unlikely to find a current word processor that reads its file format. Other formats have also become obsolete, such as those for spreadsheets and databases. Don't count on future photo programs being able to read your current JPEG photos or future music players to read your MP3 files. At some point those formats, too, will be abandoned.

In perhaps the ultimate irony of vanishing digital data, we've already lost much of the history of the World Wide Web. Many early Web sites have disappeared like extinct species. Most of the rest have changed their character over time. Surfing the Web today is a different experience than it was ten years ago. There are now several projects striving to archive the Web, but much has already been lost. And before the Web, there were almost one hundred thousand bulletin board systems (BBSs), which have also been poorly preserved. Like early radio broadcasts that existed once and are now gone, we've let an important bit of technological history slip by.

Needless to say, even the present Web is impermanent. Some sites license photos for only a limited time—creating the online equivalent of a book whose images vanish after a couple of years. Big companies have occasionally lost control of their domain names; for example, Microsoft forgot to renew passport.com in 1999 and the Washington Post forgot about washpost.com in 2004. The Web pages I cite in the notes at the end of this book *were* accurate, but they may not be by the time you read this, or the site may now require registration or a fee to view it.

Records of the early days of the PC industry are also threatened—and from more than just the temporary nature of digital data. Copyright laws forbid making copies of software. An archivist wanting to preserve, not steal, a software title may be legally prevented from doing so before it fades from its disk. And even when old software is preserved, running it requires a compatible environment. Backward compatibility (running old software) is an important attribute of most operating systems, but this extends back only so far. Titles from the dawn of the PC Age probably won't run on modern computers.

A related problem arises when a software company can't prove prior art (information in the public domain that proves who developed something first) in patent battles because their old software no longer runs on new computers. The original source code that generated that software, which might prove the point, may have been lost. Misplaced or discarded source code was also an issue with the Y2K debacle, during which companies spent billions as they scrambled to ensure that their software stored year information with four digits rather than two (1957 instead of 57) before the first two digits changed at the end of 1999. They had to find and modify the source code for software written in the 1980s and before, when saving space by using two digits to store years was prudent. With the rapid pace of computer innovation, who would have thought that this software would last so long?

A century from now, the e-mail record of Microsoft's history may be sparser than the paper correspondence documenting the rise of Standard Oil a century earlier. Our garbage may be more permanent than our written records—the writing on plastic packaging buried in most landfills will probably be more legible in a century than a newspaper or a paperback archived in a library. "Digital is forever" is a tough promise to keep. Never has the record of civilization been so impermanent.

Bugs

Technology made large populations possible;
large populations now make technology indispensable.
—Joseph Wood Krutch,
   American critic and naturalist

A software bug once credited hundreds of customers of a major U.S. bank with almost a billion dollars each. A bug caused the first launch of France's Ariane 5 rocket to explode. The Mars Climate Orbiter satellite made incorrect calculations and crashed into Mars because one part of its software used English units while another part assumed metric units. It was a bug that caused a cancer treatment machine to overexpose patients to radiation, killing several. The cruiser USS *Yorktown* was disabled in 1998, not by enemy fire, but by a computer bewildered by a request to divide by zero.

Software is brittle. Today's computers have only the most trivial ability to deal with unexpected situations, thus programmers must anticipate and prepare for them. Any surprise is a bad surprise, and an expensive project can be disabled or destroyed by a single typo.

Catastrophic bugs, like those that bring down rockets or disable navy ships, are only the most spectacular ones. Less sensational bugs can plague large software systems even before they're launched. You may have heard how bugs and the resulting cost overruns hobbled the computer-controlled baggage handling system at the Denver International Airport. A similar example is the Federal Aviation Administration's upgrade to the air traffic control system, which has been under development since the early 1980s. At the other end of the scale, bugs can affect us personally: we've all been aggravated by them in popular PC programs and operating systems.

To discover how costly bugs are, the National Institute of Standards and Technology commissioned a study in 2002 on the economic impact of software bugs. It concluded that even though software developers spend 80 percent of their time finding and fixing bugs, the cost of software bugs to the U.S. economy is $60 billion per year. That's 0.6 percent of the gross domestic product.

Software is only part of the story. Physical high-tech products can also be unreliable or simply frustrating. As consumers, we're often enchanted by new features and services, and we often ignore the imperfections or even regressions compared to earlier technologies. Telephone service over the Internet (known as Voice over Internet Protocol, or VoIP) is cheaper, but it moves telephony from the voice domain into the same data domain as the Internet—with all its susceptibility to outages, viruses, spam, and deception (such as phishing). That's quite a step backward compared to the solid reliability of the old-fashioned landline telephone—or that of other utilities like electricity, natural gas, and water. Cell phones drop conversations as you drive, and cordless phones don't work when the power is out.

The Internet can be great when it works, but service sometimes fails, Web sites go down, links get broken, and most e-mail is spam. Surveys in 2005 estimate that U.S. companies shoulder $22 billion in lost productivity because of spam and U.S. consumers spend $9 billion to fix problems caused by viruses and spyware—this after spending several billion on protection software. Many technologies are still in their infancy: grammar and spelling checks in word processors, handwriting recognition in PDAs, or relevance of sites offered in response to a Web search. Very few of the calls to the police from residential burglar alarms are valid, and the same is true for car alarms. Technological progress is jerky, with many potholes on the way.

I listened with interest as a member of NASA's Space Shuttle software team summarized how they test software. He said that the average tester would find one bug *per year*. I was working at Microsoft at the time, and the contrast was dramatic. (Let's just say that Microsoft software contains a few more bugs than that.) Software written for some products—spacecraft, airplanes, life-support systems, and appliances, for example—must be virtually bug free. Rebooting is not an option. Why isn't PC software written to that quality level?

To answer this question, let's first look at airline service quality. Imagine that you are the new head of the U.S. Federal Aviation Administration, and you are concerned about passenger frustration over delayed flights. You institute strict new policies demanding that for any flight arriving more than ten minutes later than scheduled (except for reasons out of the airlines' hands, such as bad weather), the airline pays a fine of one million dollars. Airlines could adapt to this new policy by buying more planes to use as spares in case of mechanical problems, having more spare crew members in case of problems with the scheduled crew, and dropping risky routes. Obviously, fares would increase and convenience would drop to pay for the generous safety net. It may be possible to attain this on-time nirvana, but are we willing to pay the price?

Similarly, consumer software vendors could produce much higher quality software. But just as airlines would have to adapt to stricter demands on their service, software companies would need more time and money to support this much more deliberate software creation process. If this had been the philosophy since the 1950s, programmers might be using a really, really good version of Fortran or Cobol instead of Visual C++, and PC users would have a rock-solid character-based DOS operating system instead of a Windows-based one. Not only would innovation be slow, software would be much more expensive.

Market successes suggest that while PC users would prefer more robust software, they won't trade away rapid innovation to get it. You could see bugs as *necessary*—commercial software that is bug free would simply take too long to make. In the calculus of the marketplace, bug-free software is not the better approach, and to avoid risk is the biggest risk of all.

Risks of Monoculture

Buying the right computer and getting it to work properly
is no more complicated than building a nuclear reactor
from wristwatch parts in a darkened room using only your teeth.
—DAVE BARRY, humorist

In the mid-1700s, the potato was introduced from the New World into Ireland, where it gradually became the primary food. Farmers experimented with different varieties and eventually used only the one or two highest yielding varieties nationwide. But monocultures are vulnerable, and this situation was like a pool of gasoline awaiting a spark. In 1845, a fungus made its way to Ireland, where it raced through the potato fields and destroyed much of the crop. In the following few years, this crop failure caused the century's worst famine in Europe, and it reduced the population of Ireland by about a quarter due to starvation, disease, and emigration. After the famine, Ireland's population continued to decline, and it was barely half its 1840 size at independence in 1921.

One would think that similar famines would have occurred in its homeland of Peru, where the potato is also a staple crop. Not so. Even today, Peru maintains hundreds of varieties of potato, and diseases that affect one variety often don't affect others. Widespread blight is thus avoided.

Sudden disasters have hit other industries in similar ways. In 1927 Henry Ford bought a vast area of land in the Amazon jungle and began a rubber tree plantation. He didn't understand the benefits of a natural, heterogeneous forest, and disease and insects attacked his monoculture. Nearly two decades and $10 million later, he abandoned "Fordlandia." In 1967, just twenty years later, another entrepreneur gambled on a similar plan. Daniel Ludwig, anticipating a worldwide shortage of paper, thought that the Amazon region could produce not just logs, but finished rolls of paper on a colossal scale. Everything about his project was big, from its three million acres of land, to the 2,500 miles of roads, to the seventeen-story-high preassembled pulp mill floated up the Amazon, . . . to the nearly $1 billion lost on the project.

Because PCs are typically interconnected by e-mail or networks, a computer virus or other malicious piece of software can spread quickly from PC to PC, just as a real virus spreads from person to person in an epidemic. Computer viruses are typically designed for a particular operating system. Because Microsoft Windows is the biggest target, most computer viruses are designed to attack it. This has become particularly dangerous to society because our interconnected PCs have many traits of a monoculture.

Some PC users have responded by becoming less interconnected and never opening any e-mail attachment they didn't specifically request, or even disconnecting from the Internet. Others have made themselves less of a target by moving to lower-profile operating systems. Some industry analysts have recommended that the worldwide PC environment be diversified—like the native Amazonian jungle or the Peruvian potato "multicultures" that keep plants safer from pests. However, obstacles to bad software are usually obstacles to good software. In the early days of the industry, the PC environment *was* a multiculture, but it was also a Tower of Babel in which incompatible standards slowed progress. And even then, computer viruses were a problem. The challenge is finding the balance between a vulnerable monoculture on one hand and incompatible and antagonistic fiefdoms on the other.

Technology Dependence

What's wrong with technology is that it's not connected in any real way with matters of the spirit and of the heart.
And so it does blind, ugly things quite by accident and gets hated for that.
—Robert Pirsig, *Zen and the Art of Motorcycle Maintenance* (1974)

Imagine the following science fiction scenario. You wake up one morning to what appears to be life as usual but soon realize a startling truth: everyone is gone. There's no one anywhere. All the buildings and other evidence of civilization are just as you remember them, but you're all that remains of the population. You wonder what happened to the others—maybe they were spirited away by aliens—but you don't have the luxury of speculating on the fate of your missing neighbors. You must focus on your own predicament.

At first glance, survival seems straightforward. You're in a looter's paradise, with stores full of food, clothes, and even entertainment. There's no one to stop you from taking what you need. Then a darker reality becomes apparent a few hours later when the power fails. The food in stores begins to spoil, and there is no one to replenish it. Gasoline pumps stop working. Tap water runs out. You're not sure how much longer the natural gas that heats your home will last. The future begins to look a lot more primitive.

Luckily, it would take a catastrophe to produce this situation. But the story illustrates how increasingly dependent we are on technology. The independence of the farming family is long past for most of us. Large failures are infrequent, but most of us have experienced enough small ones to appreciate the brittleness of our technology infrastructure.

Imagine, for a moment, that you are shopping during a power failure. You make it to the store past the (nonworking) traffic lights, enter through the (nonworking) automatic doors, stumble around in the dark, find a cashier willing to add up the purchases manually, and even if you did remember to bring cash, the bar code reader doesn't work. How will they know what to charge? A high-tech, centralized system to manage inventory, prices, and sales has nice advantages, but a failure of such a system highlights how dependent we really are.

Consider a simpler failure. Suppose you've recorded your friends' phone numbers into your cell phone. An integrated phone and phone book is quite handy—until you find yourself without your phone. Maybe you've misplaced it or the battery has died. Want to use mine? Unfortunately, it won't have those recorded phone numbers. Here's another example: Some car radios can display the station's call letters instead of its frequency (KUOW instead of 94.9 MHz, for example). That's a nice feature, until you want to select your favorite station on a different radio and realize that you now remember only the call letters. Computers and other intelligent gadgets are a handy place to off-load some of your memory—until that technology is gone.

Many of us off-load onto a laptop. It can hold our phone book, notebook, calendar, and other peripheral notes and reminders, plus all the files that make up one's daily business life. Everything is consolidated in one convenient place. But it's also a vulnerable place, and the convenience is forgotten when the laptop breaks, is incapacitated by a virus, or gets stolen. According to the 2002 Computer Security Institute / FBI survey, victims of the theft of a business laptop estimated their financial loss to be close to $100,000 on average (imagine walking around with your laptop case stuffed with its equivalent in cash rather than your laptop). The theft can mean the loss of confidential company data, credit card or bank information, and so on, but the clever thief may also have the passwords to access the company intranet and do even more damage. And, of course, the victim has the unwanted task of re-creating the laptop's information as completely as possible. A PDA is a similar concentration of valuable but vulnerable information.

The inevitable breakdown exposes our dependence on technology. The Y2K scare highlighted precisely this problem, and we can see it ourselves in the occasional breakdowns we experience or read about. We don't switch to an antenna when the cable TV is out. A cordless phone doesn't turn back into a landline phone when the power fails. A plane can't gracefully revert into a bus when there's trouble.

Another concern heightened by technology's progress is security. In the past, the clumsy operations of commerce were a hindrance to good guys and bad guys alike. Dealing with coins and currency demanded a lot of a bank employee's time. On the other hand, the bank robber shared that burden and could steal no more than he could carry. Now that banks can electronically move a million dollars as effortlessly as they can a single one, they are vulnerable to a hacker getting inside the electronic system and transferring money with the ease of a trusted employee. Phone security is undergoing a similar transformation. In the past, an eavesdropper had to physically tap a phone line, which was time-consuming and risky. Internet telephony means that the hacker of tomorrow may have access to phone conversations nation-wide without leaving his office.

Military hacking, where an enemy tries to destroy, damage, or alter business or military computer systems, turns hacking into a potential weapon. A society dependent on computers is vulnerable to such an attack, and the more high tech, the more vulnerable. The motto of this sort of hacker might be; "Make it Y2K day every day for your enemies . . . kick them in their electronic balls."

Human agents are only part of the problem; the natural world can be dangerous, too. Periodic meteor showers hit the Earth, which means nothing more than a light show for us, since we live beneath the atmosphere. But high above this protective layer, satellites are at risk. The occasional surges in charged particles emitted by the sun that give us nothing more dangerous than auroras also threaten satellites.

Natural cycles are often long and deceptive. We can build during a lull and only belatedly realize the danger to coastal communities during a hurricane lull, to towns on floodplains with no one to warn of the historical danger, or to cities between earthquakes or tsunamis. Recall the example of the Mississippi River in 1993, where levees encouraged development on floodplains considered safe against all but the biggest floods. Or how Hurricane Katrina in 2005 exposed New Orleans' dependence on levees. When that huge flood eventually came, huge damage resulted. We're often encouraged to trust, Pollyanna-like, in modern technology and put all our eggs in one basket. That works until the basket breaks. Increasing dependence on buggy, fragile, and brittle technology creates much of the insecurity we feel in our modern world.

> *The human race might easily permit itself*
> *to drift into a position of such dependence on the machines*
> *that it would have no practical choice*
> *but to accept all of the machines' decisions. . . .*
> *Turning them off would amount to suicide.*
> —*Unabomber Manifesto*, paragraph 173 (1995)

5 More Powerful Than a Locomotive

IN THE NEXT THREE CHAPTERS we will look carefully at nine "High-Tech Myths," each of which should sound familiar—perhaps even seductive. Each will be illustrated with claims from respected sources. Once we have this common understanding, the claims and the associated myths will be refuted.

In this chapter we look at the two most general myths that are "more powerful than a locomotive": that technology change is exponential and that technology is inevitable. But like Superman, who was also supposed to be more powerful than a locomotive, these are fiction. Raising these myths out of unconscious acceptance and exposing their flaws should help you to recognize and reject them in the future.

High-Tech Myth #1: Change Is Exponential

In the three short decades between now and the turn of the millennium,
millions of ordinary, psychologically normal people
will face an abrupt collision with the future.
—ALVIN TOFFLER, *Future Shock* (1970)

The explosive and even startling change that Moore's Law has accurately predicted for forty years directly applies only to the narrow field of semiconductors. However, many observers have seen this constantly increasing progress in other aspects of daily life. For example, economist Kenneth Boulding said in 1970, "The world of today . . . is as different from the world in which I was born as that world was from Julius Caesar's. I was born [in 1910] in the middle of human history. . . . Almost as much has happened since I was born as happened before." According to this observation, the carousel on which we are riding is spinning faster and faster.

Ray Kurzweil in *The Age of Spiritual Machines* makes an even bolder claim with his Law of Accelerating Returns, which states that the interval between important events is shrinking. He quantifies this by saying, "We're actually doubling . . . the rate of technical progress, every decade."

To further explore the implications of Moore's Law, let's compare recent change with that in the past. Let's take the period from 1810 to 1860. According to the Law of Accelerating Returns, we should now see more progress *in a week* than was made during this entire fifty-year period.

To make an even comparison, we'll compare that fifty-year period with our most recent half-century, 1950–2000. The Law of Accelerating Returns says that the progress made from 1950 to 2000 was five hundred times greater than that made from 1810 to 1860.

Of course, we're familiar with the impressive progress made from 1950 to 2000. The computer went from a laboratory curiosity, to a business necessity, to a tool on every desktop and embedded in every appliance. The Internet was created, as were cell phones. There was progress in the development of airplanes and nuclear power. Weapons became more deadly and the highway system was constructed. The entertainment industry gave us cable and satellite TV, cassette tapes, and CDs. And there was progress in many other areas: cars, plastics and chemicals, medicine, and so on.

That sounds like a lot to compete against. But look at the period from 1810 to 1860 to see what technology was doing to society at that time. You'll see that a lot more was invented than tin cans, friction matches, and safety pins.

- The telegraph went from a demonstration in 1844 to a web interconnecting most cities and bridging the Atlantic. By 1860, the world had over one hundred thousand miles of telegraph line. Information could now travel near the speed of light.

- At the beginning of this period we see Robert Fulton's nascent steamship service; at the end, we see Isambard Brunel's enormous 693-foot Great Eastern, built to carry four thousand passengers.

- The 363-mile Erie Canal was America's first manmade waterway. Completed in 1825, the canal used eighty-two locks to raise the waterway over the Allegheny Mountains. More than three thousand miles of canals were built in the United States in the following fifteen years.

- The reaper transformed agriculture—which had been a manual industry using little more than scythes, plows, and barns—and brought it into the Industrial Age. Reapers were being produced by the tens of thousands per year by the end of this period.

- The first major consumer appliance for the home was the sewing machine, also developed during this period. Sales were close to ten thousand per year in 1860 and growing rapidly. (They were over one hundred thousand per year by 1870.)

- The railroad went from an idea to a transportation revolution. About thirty thousand miles of track had been laid in the United States by 1860, with track mileage doubling every decade. (The first transcontinental railroad would be complete in 1869.)

- The manual printing press had improved little since Gutenberg's day, but that changed with the steam-driven press and later the rotary press. Volume from these machines had quickly increased over one hundred times to twenty thousand sheets per hour, driving down the cost of newspapers. This brought the penny newspaper, an explosion in the number of papers, and news-gathering services such as the Associated Press.

- The Industrial Revolution spread to America, bringing innovation and cheap goods as well as pollution and difficult working conditions.

And there were other developments during this period. Papermaking was mechanized and made much cheaper as wood fiber replaced rags. Photography was invented and refined so that by 1860 cameras were being sold to the public, portrait studios were common, photographs were printed in books and newspapers, and 3D images were popular novelties.

The idea of making interchangeable parts was perfected, first with guns and clocks, and later with other manufactured products. Machine manufacture made clocks both cheap and popular—more than half a million were sold per year by 1860. Pocket watches were also becoming affordable. Modern weapons such as rifles and revolvers were invented. Other developments during this time were the Bessemer process that produced inexpensive steel, postage stamps, pasteurization and food canning, Portland cement and reinforced concrete, vulcanized rubber, artificial fertilizer, gas lights, the ship propeller, quinine, and anesthesia.

Though it didn't bear practical fruit during this period, the foundation of computing was laid by pioneers such as Charles Babbage, Ada Lovelace, and George Boole. The period also saw early versions of the mechanical calculator, phonograph, fuel cell, fiber optic cable, fax machine, typewriter, typesetting machine, synthetic dye, light bulb, electric dynamo and motor, refrigerator, storage battery, safety elevator, microphone, internal combustion engine, oil well, airship, pneumatic rubber tire, and bicycle.

Surely no one thinks that we can duplicate this progress in a week.

Did I stack the deck by picking the half century during which the Industrial Revolution had its biggest effect? Hardly. The next half century, 1860–1910, has its own long list of fundamental inventions, including electricity production and distribution, appliances to use that electricity, the machine gun and dynamite, the early pharmaceutical industry, early plastics and synthetic fibers, affordable aluminum, sewers and clean water, the telephone, the record player, movies, and civil engineering projects including the Eiffel Tower, the first skyscrapers, the Suez Canal, and most of the Panama Canal. Also developed during this time were early versions of the car and motorcycle, the airplane, radio and vacuum tubes, color photography, and audiotape.

Let's return to Boulding's claim that as much has happened since he was born as happened before. By trivializing technology's impact before 1910, he has engaged a formidable adversary—and we've only looked back to 1810. Before that time we find inventions such as windmills and waterwheels, paper and the printing press, tools for manufacturing and farming, and the calendar and clock. The Industrial Revolution harnessed steam power and produced powered looms, spinning machines, and other machinery. Warfare progressed from arrows and spears, to cavalry and armor, to fortifications and siege weapons, to cannons and muskets.

And, of course, all this is built on a technological foundation that goes back thousands of years: architecture and the building of cities, agriculture and irrigation, domestication of plants and animals, textiles and trade, mapmaking and navigation, metallurgy and pottery, simple machines (such as levers, pulleys, and gears), civil engineering (dams, pyramids, stadiums, bridges, and so on), wheeled transportation, boats, roads, and the harnessing of fire.

(I can't help noting a few of the fundamental inventions outside of technology on which civilization is built to further put our recent progress in perspective. These include banking and money, universities and public education, encyclopedias and libraries, hospitals, cities and job specialization, art and music, mathematics and science, language and writing, organized religion, democracy, corporations, taxes and insurance, and the legal system.)

Of course we've made lots of progress in recent years, but let's not confuse its scale with the immensity of what came before. There's really no comparison.

Attack on the Exponential Model

I have seen the future, and it's still in the future.
—JACK ROSENTHAL, editor of the *New York Times Magazine*

The 180-foot-tall Home Insurance Building in Chicago, built in 1883, is often cited as the first modern skyscraper. Building heights quickly progressed through a number of records, including 309 feet in 1890 (the World Building), 792 feet in 1913 (the Woolworth Building), and 1250 feet in 1931 (the Empire State Building). After forty years of furious progress, during which record building heights increased fourfold, records did not advance *at all* for the next forty years. The current record holder, Taipei 101, is just one third taller than the Empire State Building. Clearly, an exponential curve is not a good model of this progress.

Let's look at other technologies that also have failed to maintain an exponential progression beyond a certain point. The chart of bridge lengths shows very gradual increases for a century until the George Washington Bridge in New York in 1931 doubled the previous record. Six years later, the Golden Gate Bridge added another 20 percent. But after this brief burst of innovation, almost thirty years passed before the Verrazano-Narrows Bridge exceeded the Golden Gate in length, and that was by just 1 percent. The Verrazano-Narrows is still the U.S. record holder. Dams have also shown only temporary exponential progress. The largest dam in the United States by reservoir capacity is still the Hoover dam, built in 1936. The same is true for ships: length, weight, and passenger capacity—none of these are advancing exponentially.

The chart of record airspeeds (see figure 3) starts out as a textbook exponential curve. It extends from the Wright brothers' first flight in 1903 through monoplanes, metal construction, jet engines, and the sound barrier. However, this curve ends abruptly in 1965 with the Mach 3+ flights of the SR-71 military reconnaissance plane. That was

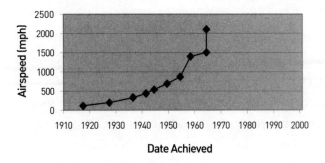

Figure 3. World record airspeeds

it—records stopped falling. In fact, the SR-71 has since been taken out of service. Want to include rocket-powered planes? The record there is Mach 6.7, set in 1967—again, with no advances since. Pilotless scramjet test planes have recently beaten this record (slightly), but we're a long way from a practical airplane using that technology. Commercial aircraft speeds have also peaked and are held below Mach 1 for economic reasons. And despite bold predictions of personal flying vehicles in the 1950s and earlier, production of new small planes dropped by 90 percent during the 1980s.

The fuel cell has been heralded as an energy solution for decades, even though it was first demonstrated back in 1839. Even if fuel cells were widely used, they wouldn't tap a revolutionary new energy source because they are just another way to use fossil fuels. (This is the twenty-first century, and we're *still* dependent on fossil fuels?) Solar cells have also been long cited as an energy solution. The photovoltaic effect that underlies their operation was first noticed over 150 years ago, but solar cells have yet to move beyond specialty applications. Batteries have also been a slow mover since their invention in 1799. More than two hundred years later, a modern AA alkaline battery holds less energy than that in a gram of sugar and costs fifteen thousand times more per energy unit than electricity from the typical local power company.

Most electric cars use lead-acid batteries, which were invented in 1859. Today, the typical electric car battery pack weighs one thousand pounds, costs $2000, must be replaced after only a few years, takes four hours to recharge, and carries the car about fifty miles per charge. By contrast, fifty miles worth of gasoline weighs fifteen pounds and can be poured in seconds. The only advantage of the electric car is that the electricity is cheaper (power plants use fossil fuels more efficiently than car engines do). Electric cars in the late 1800s could travel up to twenty miles per hour for fifty miles, not dramatically worse than today. In fact, all the early automobile speed records were set in electric cars, and an electric car was the first car to exceed sixty miles per hour, in 1899. There are obviously many areas where progress is not exponential.

Production of nuclear power was developed in the United States, but its early promise faded, leaving it as only a modest contributor of electricity. One observer noted about nuclear power, "Never in modern history has a major technology, with the full backing of industry and the government, come to such an abrupt halt."

Excusing Exponential Failures

Every airplane that significantly exceeded Mach 2
is now in a museum.
—SCOTT CROSSFIELD, X-15 test pilot

We have looked at a number of examples where the exponential model would predict exponential change that never comes about. You may, however, still object to these examples for any of the following reasons:

- *Economic Disincentives.* It is unfair to fault a slowdown in the increase of skyscraper heights if there simply is no economic justification to build taller buildings. Or airplane speeds if there is no market for faster airplanes.
- *Regulatory Disincentives.* Government restrictions can dampen growth in a particular industry.
- *Social Resistance.* A technology can be at odds with public consciousness. Nuclear power is an example.

- *Saturation.* Some technologies become saturated—for example, when enough miles of railroad track were laid. When all major hydroelectric opportunities have been exploited, no more record-setting dams can be built. And no consumer product can exceed 100 percent household penetration.

- *Failing Economy.* Growth in an industry can be tied to a boom economy, with growth slowing when the economy does.

- *Unrelated Events.* External events such as war or a depression can slow or even reverse market growth. For example, telephone penetration dropped sharply during the Great Depression. Wartime can also stimulate development, as in radar and nuclear research.

- *Technical Difficulties.* Technical obstacles are impossible to anticipate and can sidetrack progress unexpectedly. After all, you can't schedule a breakthrough; some of the problems at the forefront of engineering are just *hard.*

There are many reasons for an end to a technology's fast growth phase, and these may be totally beyond an industry's control. But note one reason *not* in this list: the time period when this takes place. Growth can be slow for many reasons, but the date is not one of them. Furious progress or obstacles to progress are no more or less likely today than a hundred years ago. Computer technology *can* change quickly now and therefore does. For various reasons, change in other areas—transportation, construction, energy, and others—is not happening quickly today. In the past, these industries had their day in the limelight; perhaps they will again. Any excuse for *why* a particular technology fails to advance exponentially is irrelevant; that technology simply becomes a counterexample to the exponential-growth paradigm.

The best evidence for the exponential model is the remarkable advancement of computer technology and growth of the Internet, but this is hardly a sufficient foundation on which to build a model that claims to apply to technology in general. The model can be sustained only by picking examples that support it and avoiding those that don't.

High-Tech Myth #2: Technology Is Inevitable

You're either part of the steamroller or part of the road.
—STEWART BRAND, commenting about
   an onrushing technology (1987)

When Andy Grove was the chairman of Intel he said, "I have a rule, one that was honed by more than thirty years in high tech. It is simple. 'What can be done, will be done.' Like a natural force, technology is impossible to hold back. It finds its way no matter what obstacles people put in its place." According to Grove, technology is unstoppable, as irresistible as gravity and as relentless as moving water.

Kurzweil has a similar viewpoint. In response to the likely end of the influence of Moore's Law over computer progress around 2020, he says: "In accordance with the Law of Accelerating Returns, another computational technology will pick up where Moore's Law will have left off, without missing a beat." He adds, "The accelerating pace of change is inexorable." As examples of silicon's possible successors he offers carbon nanotubes, DNA, and quantum computing.

Before we are swept up in this kind of optimism, we must pause to analyze the idea of inevitable change. While it's certainly true that these emerging technologies are plausible candidates to pick up where silicon leaves off, there's no guarantee that *any* will come through for us, just as Josephson junctions, bubble memories, and other computer technologies were abandoned in the 1980s despite initial high expectations. Look outside electronics: with the success of the Empire State Building in the thirties, architects projected mile-high buildings; with the success of jet airplanes, engineers began designing supersonic commercial aircraft. They guessed wrong on both counts. No technology sustains exponential progress forever, and that will also be true for computers.

Though technological failures are usually forgotten in the swirl of new product excitement, there have been plenty. By looking more closely at a few of them, we will see that success is hardly inevitable.

Federal Express launched ZapMail in the early 1980s to send mail instantly with fax machines. Businesses soon bought their own fax machines, however, and FedEx lost $300 million. Japan's Fifth Generation Computer project of the 1980s was to develop a powerful new type of computer and seize the computer initiative from the United States. It was a flop. In the early 1990s, when disillusionment with virtual reality began to show, someone noted that there were more virtual reality conferences than customers. The USS *Nautilus,* the first nuclear-powered submarine, was launched less than a decade after the first nuclear bomb test. With practical uses of nuclear power emerging so soon after the bomb, many were certain that fusion power would similarly be harnessed soon after the first hydrogen bomb in 1952—and yet we're still not even close. In 1959, a guided missile containing three thousand letters was test fired from a submarine to a Navy base in Florida. Though the letters arrived safely, some technologies—"Missile Mail" among them—are not meant to be.

In the 1950s, the hovercraft was a remarkable new vehicle. It could travel much faster than an ordinary boat and even had amphibious capabilities, but it has limited roles today. Magnetohydrodynamic (MHD) propulsion for ships was another false promise. We've seen gas turbines, rotary engines, Stirling cycle engines, flywheels, and fuel cells proposed as new sources of power for cars; magnetic levitation and monorails for trains; nuclear power for airplanes; and electricity too cheap to meter for homes. There have been backpack flying machines, autogiros and other personal flying vehicles, cars that drove themselves, moving sidewalks, and dirigibles.

You may not be familiar with all these technologies, and you needn't be. These are products that either failed completely, exist now only in niche markets, or are still not ready for prime time. The common element in all of them is that they were expected to make a much bigger and faster impact on our lives than they actually did. In fact, if you *haven't* heard of some of them, that underscores the point that these developments, a big deal in their day, climaxed with a whimper and not a bang.

Videotex trials were launched repeatedly and unsuccessfully from the late 1970s through the '80s. (Videotex systems typically displayed simple text and graphics on terminals, TVs, and eventually PCs. Though not impressive by today's standards, videotex fit well with the equipment of the day.) Other failed consumer products have been fax newspapers, videodiscs, MiniDisc, quadraphonic sound, and 3D movies. There have been TV-top CD-ROM appliances (such as the Philips CD-*i*, Commodore CDTV, and Tandy VIS), digital tape (such as DAT and DCC), AM stereo radio, and Polaroid instant photography (both still photography and home movies). And cable radio, Citizen's Band radio, and 8-track tape.

The field of artificial intelligence (AI) has repeatedly disappointed expectations. I remember the excitement I felt after reading about Terry Winograd's breakthrough AI program SHRDLU (1970). This program could be told how to manipulate blocks in a primitive graphics world *in real English.* The user could type in commands such as, "Find a block which is taller than the one you are holding and put it into the box." The program would act on the command or, if it was ambiguous, ask questions to clarify. It seemed certain that natural conversation with a computer was just around the corner. But while SHRDLU was indeed a breakthrough, it was only one step on a surprisingly long path to language understanding. *Business Week* in 1981 expressed a mainstream opinion that before the decade was out, AI advances would "have the most sweeping implications for business and society of any technology yet devised, eclipsing even the enormous changes already wrought by computers." As Ted Nelson observed about this type of overenthusiasm, "I mistook a clear view for a short distance."

We didn't need to fear the computer HAL in *2001*. Robotics and smart homes have also not met expectations, nor have videophones, biometrics, telecommuting, artificial hearts, solar power, freeze-dried food, and smart cards. We were told that nuclear bombs would be used to excavate harbors, space travel would be practical, desalinized sea water would irrigate deserts, the Network Computer would replace

PCs, and electricity would be sent without loss over cryogenic power lines. Diseases like influenza and malaria would be only memories, colonies on the moon or under the ocean would reduce the overpopulation problem, we would grow food in hydroponic greenhouses, and in a world language like Esperanto we'd marvel at mile-high buildings. We'd have paper clothes, paperless offices, hologram TV, and irradiated food.

In their day, these have all been seen as important new products, irresistible and relentless. History is littered with failed technologies once billed as inevitable. Some new technologies make it. Most don't. Based on the overabundance of new products, it seems that some companies hope that invention is the mother of necessity.

Kurzweil makes another claim that depends on an assumption of the inevitability of change: "Over the next several decades, machine competence will rival—and ultimately surpass—any particular human skill one cares to cite." In particular, he predicts that computers will pass the Turing Test by 2030.

This is likely to be one more overly optimistic artificial intelligence prediction. Speech, vision, cognition, and other elements of intelligence have proven far harder to duplicate than researchers have expected. Currently, the grammar checker in a word processor is about as advanced an example of AI as most people encounter.

The problem with the Turing Test isn't just the processor speed, as some people claim, it's the software. A faster processor alone wouldn't solve the problem because no software exists (or will soon exist) that mimics human intelligence.

Contrast this situation with the production of a computer-animated movie. We know how to make computers create each frame, even if they can't do so in real time. But this isn't a problem: the film is created over a period of months and played back in an hour or two. The analogous situation doesn't exist for Turing Test computers. We still don't know how to pass the Turing Test, so faster hardware will just give us unsatisfactory results *faster*. The track record of AI progress suggests that the missing insights may come, but not very soon.

For an example of how misunderstanding technological change can lead one down the wrong path, here is a quote from Spiro Agnew when he was vice president in 1972: "It must be obvious to anyone with any sense of history and any awareness of human nature that there *will* be SSTs. And Super SSTs. And Super-Super SSTs. Mankind is simply not going to sit back with the Boeing 747 and say 'This is as far as we go.'"

What's the harm in this sort of delusion? Quite a lot. Suppose we find a good temporary solution to the problem of nuclear waste disposal: it'll keep the waste safe for a hundred years, then we'll need to repackage it. But by then, we're thinking, technology will have advanced so far that a much better solution will have been found, right? Or energy use—can we expect new sources of energy to reduce our dependence on imported oil? Or global warming—will technology come to the rescue? Not necessarily. We're still making glass, pottery, and bricks the same basic way as the ancients did, for example, so why is a breakthrough in nuclear waste disposal—or energy, or the environment, or any field—guaranteed? Change is inevitable, but change in one particular area of technology is not.

The Greater Change Was Often in the Past

Moore's law codified our lightning speed-up in the pace
of technological change.
The acceleration of technology became exponential, officially.
—James Gleick, *Faster* (1999)

Progress happens in many ways; sometimes it's fast, sometimes it's slow, and sometimes it even regresses. Another manifestation of the erroneous expectation of exponential progress is to praise recent progress even when bigger improvements occurred in the past. This is an easy trap to fall into. For example, much has happened to improve communication in the past century. But is this even close to the amount of change brought about by the introduction of the telegraph and telephone? These nineteenth-century advancements, from physically traveling to communicate in person (or at least sending a letter over the same route) to instantaneous communication through wires across great distances, was by far the bigger improvement. For the first time, information could be sent in a nonphysical form for commercial and personal purposes.

Think of a time when communication was not as advanced. The British laws that offended the Americans and started the War of 1812 were actually repealed by Britain one day before the war started. The famous Battle of New Orleans, which made a hero of future president Andrew Jackson and was the largest British defeat for more than three hundred years, was fought more than two weeks *after* the peace treaty that ended that war was signed. Imagine when information traveled only as fast as a horse or ship could carry it, and consider the enormous changes brought about when the telegraph made communication not just faster, but *instantaneous.* Information, now no longer cargo, could be transmitted as data rather than carried. When you read the morning newspaper, you see news from around the world no more than twenty-four hours old—which is exactly what the telegraph brought to the newspaper of more than a century ago.

Here's another example. Compare the improvements in civil engineering from 1940 to 2000 to those in the sixty years before 1940 when concrete and steel replaced stone and wood, and machinery replaced muscle. Remember that the Panama Canal, Empire State Building, Hoover Dam, and Golden Gate Bridge were all built before 1940. Or, medical treatments and equipment developed since the mid-twentieth century versus X-rays, anesthesia, vaccines, antibiotics, antiseptic practices, and surgical techniques developed in the period before. Or the change in books from 1955 to 2000 versus the same time span five hundred years earlier, the period beginning with Gutenberg's Bible and ending with a thousand printing shops throughout Europe and millions of newly printed books.

These are examples of technological progress slowing down. Only by picking supportive examples can one make a superficial case that technological progress is exponential. These examples make a strong case that the big advance was often made in the past.

Some of technology's proudest achievements have become embarrassing examples of retreats. Most of us have heard the lament, "If they can put a man on the moon, why can't they . . . ?" with the missing challenge being anything from curing cancer to making some decent pastrami. The irony is that they can *no longer* put a man on the moon. They could in the early 1970s, but that capability has since dissolved. In three years, twelve astronauts were put on the moon. Since 1972, no one has been back; in fact, no one has even left Earth's orbit. Technology doesn't just surge forward, immune to all limit or setback.

In 2004 the Bush administration suggested a new focus for NASA: they would put astronauts on the Moon by 2020 and then, building on this experience, get to Mars. But why should NASA need sixteen years to get astronauts to the Moon now when it took them half that time in the 1960s? This is especially puzzling when we have much better technology *and we know how to do it.*

At about the time of the Apollo program, supersonic passenger planes seemed a logical step in the progression of faster passenger travel. A consortium of European companies accepted the challenge and built the Concorde. Launched in 1976, it halved the New York–Paris flight time. Estimates predicted four hundred Concordes by 1980, but only fourteen ever entered service. In 2003 the plane was retired without any successor; there was not even one on the drawing board. By the centennial of the Wright brothers' first flight, supersonic passenger aircraft had stopped flying.

Not only is the longevity of risky new programs uncertain, but the very durability of our own technology isn't what you might expect. With proper maintenance, the Golden Gate Bridge is expected to last a total of about 250 years. And yet, Chartres Cathedral is *already* 800 years old—and it was built with medieval engineering. Not only does the permanence of Chartres Cathedral outshine that of many of our modern projects, the speed with which it was built mocks equivalent projects today.

There are many instances where the big jump was a recent one: computing power, wireless communication, entertainment technologies, and others. However, don't let the recent advances overshadow advances from the past, which were often much more important. There may indeed have been a big recent improvement, but look at the entire history of a technology to place it in perspective. The landscape of technology has examples of phenomenal progress mixed with frustrating stagnation.

> *I have always wished*
> *that my computer would be as easy to use as my telephone.*
> *My wish has come true.*
> *I no longer know how to use my telephone.*
> —BJARNE STROUSTRUP, computer science professor and
> designer of C++ programming language

6 Faster Than a Speeding Bullet

IN THIS CHAPTER WE WILL LOOK at the *faster* myths—five High-Tech Myths that lead us to believe that things are happening at an ever-increasing pace. Though important to understand in their own right, exploring these myths can also help us explore one of the problems that causes these distortions: press overeagerness. When it comes to new products, what we know is largely what we hear and read from the press, and this information can be less than accurate.

High-Tech Myth #3:
Important New Products Arrive Ever Faster

The only constant is change.
—ANONYMOUS

A 1998 article in *Forbes* magazine claimed that, "new inventions now arrive at a bewildering rate—as many in a year as once appeared in a millennium." The Federal Reserve Bank of Dallas in its 1996 annual report observed, "More than half of U.S. patents have been issued in the past forty years [though the Patent Office has been open for more than two hundred years]. The number of new products put on the market annually has tripled since 1980, and with so much R & D occurring, companies are likely to keep offering innovative goods and services at a furious pace."

The *Forbes* article argues its case by listing important inventions throughout history. Inventions become increasingly numerous as time progresses, which is taken to imply a rapidly increasing rate of invention. The present-day end of the list includes important inventions such as the microprocessor, fiber optics, the Internet, and the copy machine. No one will deny that these are important. But the list also includes less important inventions such as the cordless phone and stereolithography, dubious ones such as Post-It notes and spandex, and ones that are actually still in development like the instant language translator (reportedly available in 1992) and computer speech recognition (1994). In addition, the beginning of the list omits important developments such as pyramids and cathedrals, bronze and steel, language, writing, domesticated animals, and money. Because of these omissions and because fundamental inventions such as paper and the alphabet are mixed with less important inventions, it becomes harder to see through mere *numbers* of inventions to an appreciation of their *impact* over time.

The Census Bureau confirms that the number of new products introduced in the United States went up three to four times from 1980 to 1997 for the categories of food, beverages, and health and beauty. A sign of increased innovation? Not really—most were simply new formulations, new market positioning, or new packaging. Only a tiny fraction was new because of a new technology. The per capita rate of patent introduction may actually be *dropping*. We are seeing a burst of marketing enthusiasm, not of technological innovation. Of the top twenty-five brands of the 1920s (Kodak, Coca Cola, Campbell's, and so on), nineteen were still number one in their category sixty years later. New consumer products, unfortunately, are too often like weak movie sequels.

High-Tech Myth #4:
The Rising Tide of Valuable Information

We are drowning in information
and starved for knowledge.
—JOHN NAISBITT, *Megatrends* (1982)

According to Richard Saul Wurman in his book *Information Anxiety*, "A weekday edition of the *New York Times* contains more information than the average person was likely to come across in a lifetime in seventeenth-century England." Peter Large elaborates: "More new information [has been produced in the last] thirty years than in the previous 5,000. . . . The total of all printed knowledge is doubling every eight years." Apparently, we're drowning in a deluge of information.

Wurman's quote is certainly colorful. While it would be very difficult to prove, let's assume that it is correct. A weekday edition of the *New York Times* might also hold more information than the Bible or the collected works of Shakespeare, but so what? This quote ignores the difference in importance between low-level *information* (news, facts, and other types of data) and higher-level *knowledge* (selected information particularly relevant to its owner). There is a world of difference between a newspaper's classified ads section and the job and life skills, sense of culture, ethics, and wisdom passed on from a seventeenth-century parent to a child. And if the point of this quote is the *newness* of this geyser of information, note that steam presses enabled the production of inexpensive, fast newspapers well over 150 years ago.

Compare a cheap novel with Homer's *Iliad* or Shakespeare's *Hamlet*. Compare a randomly chosen nonfiction book with Newton's *Principia Mathematica* or Einstein's collected papers. Compare the value of last Tuesday's news, now sitting in the recycling bin, to that of a compelling quote or a beloved poem or a famous speech or marriage vows or the transistor patent or the U.S. Constitution. The amount of information generated per year may be increasing, but mere information is very different from knowledge or wisdom. How much is all the information

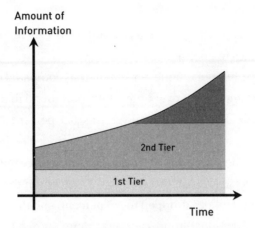

Figure 4. When the information channel is narrow, only the most important information is sent; when it is ample, there is room for more trivial information.

in that *New York Times* worth? It's worth what you pay for it—about a dollar.

The value of information spread by new technologies is illustrated in figure 4. When spreading information was expensive, only the most important was passed on. As costs dropped, more and more information of less and less importance made the cut. The most important information—at least for the past few centuries—has *always* had an outlet. Information that is new to us and that didn't have an outlet in the past, didn't have an outlet in the past *for a reason*. The newest categories of information are the least important.

High-Tech Myth #5:
Today's High-Tech Price Reductions Are Unprecedented

Why do we need another airplane? We already have one.
—ANONYMOUS CONGRESSMAN, referring to the 1908
   airplane purchased by the Army Signal Corps

Consumers expect prices for essentials such as food, cars, clothing, and housing to increase over time. A few categories such as health care and college tuition increase even faster than inflation. Yet we have the marvelous exception of PC power dramatically increasing while its price continues to decrease.

Impressive, but not unique. A 1997 analysis compared the changing costs of consumer goods over time. The report found that the cost of a given amount of computing power was 0.6 percent (adjusted for inflation) of what it was in 1984, and it has continued to drop since. But PCs were not the only products with significant price reductions. The price of a color television dropped to 4.1 percent of that at its introduction, the calculator dropped to 2.5 percent, and the refrigerator to 2.2 percent. Even more impressive, electricity dropped to 0.5 percent and a coast-to-coast telephone call to 0.04 percent.

These historic comparisons are not fair. The improvement in telephone service is understated: today's long-distance telephone call is not only vastly cheaper but can be placed quickly and is clear and reliable—unlike the 1915 version. The quality of the current generation of TVs, calculators, and refrigerators is also much improved and many features have been added over early models; they're not just cheaper.

By contrast, the improvement in computers is substantially overstated. The cost of *computing power* (dollars per million instructions per second, for example) has indeed dropped dramatically, but the cost of a *computer* has only dropped moderately. A PC today might be a quarter of the price of an original IBM PC—a nice drop over twenty-five years, but nothing unique. And it gets worse. Once we add the ongoing costs of software, training, support, and upgrades, it's clear that the cost of being a PC owner is much more than simply the cost of a single PC. More importantly, a PC with one hundred times more computing power is not one hundred times more useful for typical desktop applications. In fact, it may not even be *two* times more useful for fundamental applications like word processing or e-mail.

Let's look at why the cost of PCs doesn't drop exponentially, even though the cost of a given amount of computing horsepower does. In accordance with Moore's Law, the number of transistors on a silicon wafer has doubled every two years or so, halving the cost per transistor but keeping the cost of the microprocessor roughly constant. Even if Moore's Law were applied to produce cheaper microprocessors of constant computing power, the cost of a PC wouldn't drop proportionately because the microprocessor is a small fraction of the total cost of a PC.

High-Tech Myth #6: Products Are Adopted Faster

Web time [is] seven times faster than normal time.
—*Cluetrain Manifesto* (1999)

Another popular myth states that products are reaching us at an increasing rate. The U.S. Department of Commerce outlined the argument this way: "Radio was in existence 38 years before 50 million people tuned in; TV took 13 years to reach that benchmark. Sixteen years after the first PC kit came out, 50 million people were using one. Once it was opened to the general public, the Internet crossed that line in four years."

This is hardly a fair comparison. Fifty million was half the U.S. population when radio was introduced, but only 20 percent when the Web started. The "once it was open to the general public" caveat for the Internet is also important. The Internet began in 1969. This means that twenty-two years of money and research from the government and universities nurtured it before it was opened to the public in 1991. And even at its starting point in 1969, the Internet wasn't built from scratch, like radio or the telegraph, but on the infrastructure and experience of the telephone industry.

This is rather like a bamboo plant that builds its root infrastructure for years and then bursts forth with a new shoot that grows a foot or

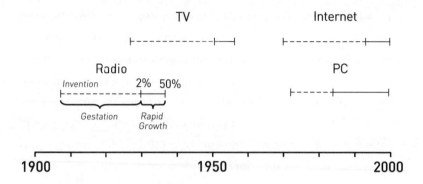

Figure 5. Four important technologies of the past century took similar amounts of time to mature.

more per day. It can be said that the bamboo grows to full height in a month, but that ignores the years of preparation that made it possible. Not only was the Internet nurtured for decades before the Web was introduced, but by the time it was made available to the public, the home PC industry was already well established. From the consumer standpoint, the Internet was born with the technological equivalent of a silver spoon in its mouth.

This quote could be more honestly written as follows (see figure 5).

The first radio broadcast was in 1906. About twenty-three years later, radio was mature enough for consumer use and receivers were in 2 percent of American households. Radio was in 50 percent of households in seven more years. Television, invented two decades later, had a similar progression: twenty-four years to reach 2 percent penetration and six more years to reach 50 percent.

If we take the year the first microprocessor was built (1971) as the start of the PC industry, it took little more than a decade to reach 2 percent of the population. Its gestation was much faster than that of radio and TV because the PC did not need as much infrastructure. Nevertheless, it took almost *two more decades* for PCs to reach 50 percent penetration, three times longer than radio or TV.

The Internet was begun in 1969 as a government-funded research project. It was opened to commercial use the same year the Web was launched, in 1991. It took twenty-four years to reach 2 percent household penetration, and it hit 50 percent only after an additional seven years.

What conclusions can we draw? The evidence for accelerating technology change has evaporated, and we can see that successful products throughout the past century had similar gestation times and growth rates, and that modern inventions have not reached the market unusually quickly. It's also interesting to note that the PC, one of the poster children of the our-times-are-unprecedented mindset, grew so much more slowly than radio and TV. Don't think that the PC carried a heavier burden because it was expensive. A 1981 PC was half the relative cost of a 1939 television and one tenth that of a 1908 Model T (these three dates are the first time their respective products were made available to the general public).

We'll take a final look at this question of how fast technology moves by returning to the example of cathedral construction in the Middle Ages. Imagine a cavernous, handmade stone building over four hundred feet long with twelve stories of open space inside. The primitive cement of the time does little more than fill the gaps between the stones and will break if tension (pulling force) develops. Builders can test new techniques and designs only through experimentation. Most are illiterate; they learned their skills through apprenticeship rather than books or schools. Architecture is not yet a science, and failed experiments can cost lives and years of work. There are no cranes, trucks, or power tools—there are not even any blueprints. This was the challenge facing the town of Chartres, France, in 1194.

Despite these difficulties, the stunning Chartres Cathedral, which still stands today, was almost completely built within thirty years. Salisbury Cathedral in England, of similar dimensions and begun a few decades later, took less than forty years to complete. It is humbling to note that Washington's National Cathedral took more than twice this long, and New York City's Cathedral of St. John is still unfinished after over a century of work. The common perception of medieval cathedrals requiring centuries of work from generation after generation of stonemasons is quaint but not always true. When funds were available, as they were for Chartres and Salisbury, work proceeded quickly. When they are not, as in these modern examples, work halts.

High-Tech Myth #7:
Invention Gestation Time Is Decreasing

Despite the awe that many express about
today's technological developments,
the material innovations in our everyday lives are incremental
compared to those around the turn of the [twentieth] century. . . .
Today's technical whirl is by comparison merely a slow waltz.
—CLAUDE FISCHER, *America Calling* (1992)

The previous section examined how fast a product sweeps over us once it reaches the marketplace. Now, let's look at the previous phase: from the bright idea or laboratory prototype to the viable product poised to take off.

Inventions take longer to reach the consumer than you may think. Too often we remember only the success of a product's latest incarnation and ignore the years of failure or weak sales that preceded it. Today's overnight sensation likely endured decades of now-forgotten experiments, field trials, and disappointments before becoming a popular product. For centuries, inventions have typically taken twenty to thirty years to progress from the first patent or prototype to a widely used product.

The Xerox Star was the first computer to ship with a mouse in 1981 (three years before the Apple Macintosh), but the mouse was invented in 1963. VCRs moved into homes quickly after the 1975 introduction of Sony Betamax, but that was almost thirty years after the first videotape demonstration. The bar code (UPC) was standardized in 1973, over twenty years after the first bar code patent application. The answering machine that became popular in the 1970s was not the first of its kind—primitive versions preceded it by decades, as had answering services. Microwave ovens were available soon after the discovery of microwaves' heating potential, but twenty years passed before the first successful home microwave oven was put on the market in 1967.

The examples continue. The time from the first magnetic tape recorder to the introduction of the audiocassette (in 1961) was twenty-six years; from copier patent to the first Xerox copier (1960), twenty-three years; from the design of the modern television to regular U.S. broadcasts (1939), seventeen years; from the first long-distance radio broadcast to the first commercial radio station (1920), nineteen years; from Nikolaus Otto's gasoline engine to the Ford Model A (1903), twenty-seven years; from design to the first gramophone for the home (1896), nineteen years; from the invention of the telegraph until more than twenty thousand miles of line were in place (1852), fifteen years; from the first section of railroad track in the United States until six thousand miles of track were in place (1848), eighteen years.

Even the World Wide Web had precedents. The first popular browser, Mosaic, was introduced in 1993 and made the Web conveniently accessible to the public. But the Web was not a complete novelty. By that date, France's Minitel information and communication service had been in use for a decade and already had six million users. The United States had tens of thousands of electronic bulletin boards, accessible by any PC with a modem. We had also seen more than a dozen high-profile videotex trials since the late 1970s in the United States alone. These trials were funded by heavyweights such as American Express, AT&T, Knight-Ridder, Dow Jones, Time, CBS, Sears, and IBM. Though the initial forms of these videotex services have long since been discarded, this period of innovation did yield important lessons and created CompuServe (1979), Prodigy (1984), and America Online (1985).

The Web, running on the Internet, is the technology that caught on, but vast amounts of work preceded its introduction. It caught on as fast as it did because more than two decades of press about the value of information services prepared the market. Just like the rest, the Web wasn't an overnight success.

Press Overhype

Were it left to me to decide whether we should have
a government without newspapers,
or newspapers without a government,
I should not hesitate a moment to prefer the latter.
—THOMAS JEFFERSON

Public concern about violent crime increased substantially around 1990, not because violent crime was increasing, but because *press coverage* of it was. In a similar way, the press has been a powerful force steering us toward our current view of technology. Enthusiastic stories about developments still in the lab blur the distinction between the potential product and the successful product, even though any entrepreneur can tell you that there's a big difference. A related problem comes from companies who predict greatness for their own new technology. They hope that enthusiasm in the press will turn into investment or sales, and their predictions are intentionally self-serving.

The press treats new technologies like a slot machine treats a gambler: when a machine pays a jackpot, bells ring, lights flash, and the coins clink noisily onto a metal tray. But when a player doesn't win, the loss happens anonymously and silently. The bystander feels surrounded by winners, though winners are actually a small minority. In the field of emerging technology, the media is that noisy slot machine, celebrating the success of each new technology. In this casino, everyone's a winner! Failed technologies—either those that never leave the lab or those that bomb in the marketplace—are usually ignored. There may eventually be a "Whatever happened to . . ." article, but any correction is tiny compared to the overhype that preceded the fall. It's like the psychic who boasts of his few successful predictions and hopes everyone forgets his many failures. And in the case of the press, that hope is well placed: perhaps something innate in all of us cheers for the up-and-coming new development.

The press buoys our enthusiasm for new products like a parent encouraging a child to swim. "Swim to me," the parent says. "Look—I'm not far away." As the child swims, the parent may pull back, giving the child a longer distance, to show that he can make it farther than he thought. The press unintentionally plays the same trick. "Here's a great new technology! It's not far away." By the time we actually get there—*if* we get there—the wait was usually much longer than promised. While the deception may have been innocent, we must learn from it and avoid being fooled in the future.

We can anticipate how the press will treat future developments. Perhaps by 2010 HDTV will be at the point on its growth curve that television reached in about 1950, when household penetration exceeded a few percent and sales began to accelerate. When the press writes about the impressive HDTV sales figures, there will probably be much marveling at the speed with which it is taking off. But will there be comparisons with television's phenomenal growth in the fifties or radio's in the twenties? Will it be noted that HDTV has been demonstrated since the early 1980s? HDTV will be an overnight success, thirty years in the making. To take another example, if video telephony finally succeeds—over the Internet, perhaps—will the public also see this as a new product and ignore the videophone's struggles since 1964 and before?

Superficial press plus technological myopia cause us to see today's change as more important than it really is while simultaneously minimizing the significance of past technologies. The result is that society sees technology from a warped vantage point. The comparison of today's developments against those of the past is not fair. On one hand we have marching bands trumpeting today's shiny new technology plus those exciting products-to-be from the cover of *Popular Science*. On the other, we have those few old and familiar technologies that manage to break through the mist of time—hardly an equal comparison. When looking at the products that affect our daily lives, it is difficult to see beyond the new veneer to appreciate the substantial foundation laid during previous centuries.

To judge by the overheated tributes to computer technology
that have become increasingly common in the press . . .
one would be led to conclude that the Internet
is the most important invention since fire, [and]
that a laptop computer dwarfs the automobile in its societal impact. . . .
My advice to the somewhat overly enthusiastic technophiles . . . is simple:
Get a grip.

—MICHAEL HAMMER

7 Leap Tall Buildings in a Single Bound

THE FINAL TWO HIGH-TECH MYTHS are about technologies that most of us use frequently—the Internet and PC. Because they have been evolving right in front of our eyes, we have been able to witness all the commentary that positions them prominently in the pantheon of technology milestones. Much of this, however, has been inflated. The Internet and PC are important enough in their own right that we don't need to puff them up with false importance or giddily treat them as celebrities. Only by seeing these two essential developments for what they are can we adopt them when they add value and ignore them when the benefit doesn't outweigh the cost.

High-Tech Myth #8: The Internet Changes Everything

[The Internet is] the most transforming technological event
since the capture of fire.
I used to think that it was just the biggest thing since Gutenberg,
but now I think you have to go back farther.
—JOHN PERRY BARLOW, an Electronic
   Frontier Foundation founder (1995)

Just how important is the Internet? Is it the biggest thing since fire? Does it change everything? It is obviously an important development, but to evaluate it accurately, we need to put it in a proper context.

The Internet provides an information outlet that wasn't available before. As we discussed in Myth #4 above, however, the most important information has always had an outlet, and the newest information is the least important. Additionally, much copyrighted information is either unavailable over the Internet or requires a fee.

The Internet is a big technological advance, but we can't ignore the huge progress in communication technologies before the Internet. The printing press revolutionized the copying of information, and the public library was a technologically unsophisticated but monumental advance in providing access to that information. Encyclopedias, almanacs, magazines, newspapers, and so on have been available for centuries. Combine that with the cheaper printing enabled by fast presses in the 1820s and machine typesetting in the 1880s, then printed photography, then cheap paper, and then paperback books (plus the telegraph and telephone to collect this information), and we see that the Internet is not the first, or even the most important, development in communication technology.

The Internet is a gateway to a flood of information, but it's a flood of decreasing reliability. Books, magazines, and encyclopedias go through a lot of review by editors and peers before the public sees them. The beauty of the Internet is that anyone can have a voice, but this can be at the expense of any review process. When we let a thousand flowers bloom, we get many dandelions.

The Internet gives us news that's recent and varied, but the problem that this addresses—relatively homogenized news, with just one or two local newspapers and a few papers with national reach like the *New York Times* or *USA Today*—is a recent one. Newspapers have been much more diverse in the past. The number of American newspapers peaked around 1900 with 2,600 dailies and 14,000 weeklies. There were papers for every segment of the population: immigrants, socialists, farmers, business leaders, various ethnic groups, and so on. Before that, the penny newspaper in the 1830s and the telegraph network in the 1860s put yesterday's events from around the world into this morning's newspaper and made news affordable and accessible to the average person. Note also that Internet news must still be acquired the old-fashioned way: a journalist must first hear about a story and either travel to the news site or telephone someone to conduct an interview. Next, the story must be typed and posted. To see a revolution in the delivery of breaking news, look to radio and TV, not the Internet.

The Internet (in the form of the Web) invaded our lives quickly, but we should understand the subsidy it was given. To see how the Web grew so much faster than the telephone, imagine two interstate highway systems. The first is the U.S. highway system as it was actually constructed in the 1950s and '60s. It was built gradually, and segments were opened to the public as soon as they were ready. This is how the telephone system developed. It was built from scratch with technology invented as needed. Customers had to be educated. Why should I sign up for phone service, they would demand, when hardly anyone else has done so?

Let's imagine another highway system. Here, the highways are built at the same pace but are used exclusively for government and commerce, not by individuals. Then, after a couple of decades, the completed network is thrown open to the public. The highway system is used by rapidly increasing numbers of people, businesses quickly spring up to serve them, and it quickly becomes an essential part of the society's infrastructure. This second approach was how the Web was introduced to the world—with the skeleton of the Internet infrastructure already complete and PCs in more than 20 percent of homes.

Technologies with high infrastructure needs—cars, electricity, telephone, and the Internet on which the Web runs—advance more slowly and will continue to do so in the future. Some Internet applications are important, such as e-mail, research, company Web sites, and e-commerce. Some are new, such as connecting members of obscure hobbies or finding buyers for used goods. But the important applications aren't new and the new ones aren't important.

Sense from Statistics

The throttle has been pushed so far forward in recent years that
"No exaggeration, no hyperbole, no outrage
can realistically describe the extent and pace of change. . . .
In fact, only the exaggerations appear to be true."
—WARREN BENNIS, quoted in
  Future Shock (1970)

The press is full of statistics about the fast growth of the Internet. For example, the Internet industry has been credited with roughly $800 billion in annual revenues, versus $350 billion for the auto industry and $225 billion for energy.

Can the Internet possibly be more important than energy or cars? Think about the fallback each industry has and ask yourself: If society had to lose every vestige of the energy industry (electricity, gasoline, diesel and jet fuel, heating oil, and so on) or the Internet, which would hurt more? It seems clear that energy is the more fundamental utility.

Another problem is that these annual revenue numbers aren't comparable. To see this, imagine two purchase situations. In situation 1, the consumer pays $100 to a company. The company declares $100 in revenue—pretty simple.

In situation 2 (see figure 6), the consumer pays the same $100 to the same company for the same product. But now we'll look inside this industry to see how the money flows. Company 1 declares $100 in revenue. It keeps $10 for its own operation and profit but must pay $90 to Company 2 for materials it needs to make that product. Company 2 keeps $10 and pays $80 to Company 3, and so on, down to Company 10.

The two transactions are identical from the consumer's standpoint, but by looking inside the industry in situation 2, we can tally the revenues of all the companies. Somehow the consumer's $100 has mushroomed into $550 in aggregate revenue. Is this situation five and a half times bigger or more important to society in *any* way than the first? If we fragment the industry into even smaller companies so that we can

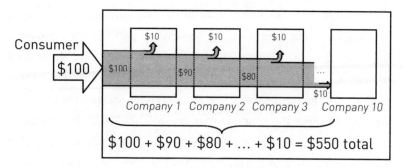

Figure 6. Mismeasurement of revenue from a single transaction

count even more "revenue," would that make the industry more important still? Not at all. Situations 1 and 2 are the *same* situation. The meaningless exercise of adding up various companies' revenues only obscures the simple $100 transaction.

Let's return to the auto versus Internet industry example. The $350 billion for the U.S. auto industry is the price paid for all new vehicles in that year. That's $350 billion paid from *outside* the auto industry *into* the industry—pretty simple. Contrast this with the approach used to measure the Internet industry. The $800 billion cited as the size of revenues for this industry includes costs for servers, fiber optics, software, consultants, and other infrastructure costs that are internal to the industry and *not* paid for directly by consumers. A small fraction of the total is online travel agents, e-commerce, and subscription services, which *are* paid from outside the Internet industry. To make a meaningful comparison, the $350 billion paid into the car industry should be compared to the corresponding amount paid into the Internet industry—perhaps only a third of the inflated $800 billion presumed in the quote above.

We also shouldn't get overexcited about the volume of e-commerce. E-commerce was responsible for about $70 billion in sales in 2004. But of total non-store sales (sales made by phone, mail, or the Internet), e-commerce accounted for just 20 percent. And of the $3.6 trillion of total retail sales, e-commerce accounted for just *2* percent. Even these small numbers are suspect. In what fraction of e-commerce sales was the real work done by a catalog or TV ad, with the consumer simply choosing to fill out a form online rather than perform the equivalent task by phone or mail? Is it fair to credit e-commerce for these sales when an old-fashioned medium did the heavy lifting? Admittedly, these numbers are changing rapidly and you may have access to more recent ones, but I believe the general argument will hold true for some time.

The final problem with the Internet industry revenue figures is that out of the small fraction that comes from outside the industry, a large portion is simply diverted from other industries. There's a big difference between a *new* sale (the mail-order catalog and railroad allowed rural families to buy things they never could before) and a *diverted* sale (a purchase over the Internet is usually one less purchase from a store, not an additional sale).

Since the late 1880s, catalogs have enabled new sales in a new way: consumers could purchase goods *without going to the store*. What do Internet sales do? They allow you to purchase goods . . . um . . . without going to the store. The product arrives in the same way and takes the same amount of time whether purchased online or ordered by telephone from a catalog. For nineteenth-century rural residents unable to get to big stores, catalogs were a revolutionary new way to buy. Contrast this period with the early 1990s when e-commerce made its appearance: transportation was convenient, stores were everywhere, and selection was excellent—*and* catalog sales meant that phoning in an order was an alternative to driving to the store. E-commerce has been an important addition but hardly a revolutionary one.

The Internet—the Plastic of the Twenty-first Century?

Most of the content on the Net is total garbage.
—ESTHER DYSON

For over thirty years, from its birth through the 1970s, the modern plastics industry grew exponentially. Today it produces over two hundred billion pounds of plastic each year. Plastic is ubiquitous. Sometimes it is put to important uses (lighter and more fuel-efficient cars, cheaper consumer products), but often its utility is of questionable importance (packaging, toys, pink flamingos). There are a number of products that couldn't be built without plastic, such as disposable medical components, but usually plastic is simply a cheaper and more durable alternative to cloth, wood, metal, rubber, glass, or other materials.

Compare this to the Internet. It is ubiquitous, or becoming so. Sometimes it offers an important service (e-mail, company Web sites, online research) but this often isn't so (personal home pages, most blogs). A few capabilities couldn't exist without the Internet, such as eBay, but in most cases the Internet is simply a more convenient alternative to existing facilities like libraries and stores. Plastic and the Internet—is there a resemblance?

Recently, a company adopted an advertising slogan that cleverly puts its products in perspective. BASF, a producer of plastics and other materials, claims that, "We don't make a lot of the products you buy. We make a lot of the products you buy better." The ad notes that they don't make the mattress, they make it softer; they don't make the boat, they make it faster; they don't make the sunscreen, they make it stronger; and so on. This is like the Internet—it doesn't create the ability to communicate or discover or shop, but rather lets you do those things better.

Accurately seeing the value through the hype is difficult. We were told in the 1990s that, "the Internet changes everything." It doesn't. That shouldn't surprise us because we were also told in the 1950s that nuclear power changed everything and in the 1960s that space technology changed everything. *Consumer Reports* commented on this change in how we see the Internet in its 2005 "State of the Net" survey: "The Internet is no longer the urbane information motorway it was five years ago. It's more like a no-holds-barred raceway teeming with unsavory drivers and with hardly a police car in sight."

The Internet is impressive enough without inflating its value. In fact, we must learn to assess it accurately to avoid a backlash when reality doesn't live up to expectations.

High-Tech Myth #9: Moore's Law Really Matters

The computer is the most extraordinary of man's technological clothing;
it's an extension of our central nervous system.
Beside it, the wheel is a mere hula-hoop.
—MARSHALL MCLUHAN

Moore's Law, impressive though it is, usually yields fewer benefits than one might expect. Exponential growth in the power of computers is sometimes confronted with exponential demands. For example, suppose that displaying text and making simple tones takes one unit of computing power. Adding audio and color images as the next level might take ten times as much power. Adding support for video playback and limited animation might demand another such increase. Many more order-of-magnitude jumps might make it possible to produce real-time animation of a movie like *Toy Story.* Progressively more power-hungry applications gobble power at an exponential rate, but the net effect is linear improvement for the consumer. A picture may be worth a thousand words, but it demands the memory, bandwidth, and computing power of *ten* thousand words.

A video game that uses ten times more computing power will not have ten times as many users or get ten times as much use. Videophones have struggled with this problem—they need much more bandwidth and computing power than telephones but return little additional benefit. In short, the applications that need the exponential performance increases often have exponential demands: the next small step up the ladder takes a *lot* more computing horsepower.

This phenomenon of diminishing returns can be seen in other areas. The development time for a product can be cut in half, but this doesn't double benefits to consumers. And if a PC responds to a user's click in 0.02 seconds, halving the response time to 0.01 second with a faster PC is pointless because the response was already perceived as instantaneous. Doubling a car's horsepower will not double its top speed, and even with enough additional horsepower to double the top speed, the car may not get you from your house to the store any faster, given the realities of traffic, stoplights, and speed limits.

We see the same thing in other fields, where an impressive gain in one area doesn't look as impressive from the perspective of the entire system. For example, the shipping industry focused on building larger, faster, and more economical ships for years, forgetting that the transit time between ports is only part of the cost of shipping. The narrowness of this approach was highlighted with the success of containerized shipping. When this innovation was introduced around 1960, the loading and unloading time was dramatically cut, and the time in port became less of a limiting factor. Another example is the Concorde, which halved the flight time between New York and London. Impressive, but no passenger is interested in just airport-to-airport travel time. An entire trip includes time in transit to the airport, the time at the airport checking in and waiting for the flight, and the transit time to the final destination—none of which the Concorde can improve. The result was an impressive innovation that didn't cut the origin-to-destination time by much.

Moore's Law doubles processor performance every two years, but who does this benefit? Some fields that use computers need all the processor speed they can get, such as engineering (car-crash simulations), entertainment (movie special effects), and science (weather forecasting). Except for games, this is rarely the case for consumer applications. Consider this: for a writer needing only the basic features of a word processor, the last two decades of PC improvement have not helped *at all*. True, the niceties of pictures, pretty fonts, and spell checking were not available at the dawn of the PC era, but for the fundamental operations of entering, editing, moving, and saving text, an original IBM PC with a 6 MHz processor gets the job done as quickly as a PC with a processor a thousand times as powerful.

"Ellen builds a sand pyramid one meter tall. Tom builds a pyramid two meters tall. How much sand does Tom need compared to Ellen?" The answer is *eight times* as much. It takes a lot of improvement in the foundation to make a taller pyramid, and the same is true for the improvement of the fundamental specifications of a computer required to make a noticeable improvement at the user's level. Doubling PC speed rarely doubles any metric the user cares about. We must remain skeptical of impressive low-level specifications and instead look for numbers that mean something to the typical user.

Industry observers focus primarily on Moore's Law when predicting the future of the computer industry. They wonder how long the technical progress will continue, but that may be the wrong concern. Other industries have slowed or retrenched for lots of reasons—is the PC industry immune to all of them? Maybe people will conclude that faster PCs aren't worth the expense and will reduce their demand. Consumers and businesses might ask themselves: "Why am I replacing a working two-year-old PC with a new one? It gets the job done. Couldn't I just keep this one for three years? Or five? Or eight?" Maybe only serious game players will need frequent hardware upgrades. Maybe software producers will refocus their energies on making existing hardware more productive. Even now, some PC makers are redirecting Moore's Law, not to make ever-faster processors but to make them cheaper or less power-hungry.

Gordon Moore proposed another law known as Moore's Second Law. Less well known, this law may create the toughest obstacle to PCs' continued speedup. It predicts another exponential curve: that the cost of new integrated circuit fabrication plants doubles every four years. The increase in factory costs has been supported by the increase in demand for their products, but if markets saturate, demand may drop and Moore's Second Law may deflate Moore's First Law.

Computers in Schools

What's wrong with education
cannot be fixed with technology.
—STEVE JOBS, cofounder of
  Apple Computer, Inc.

Seymour Papert, an MIT computer scientist, illustrates how little technology has helped education with the following example. Imagine that a doctor and a teacher were transported from a century ago to the present. Technology has so changed today's medical landscape, with new tests, drugs, knowledge, techniques, and equipment, that the doctor would be unable to practice medicine. Nevertheless, beyond a few small adjustments, a teacher from a century ago would fit well into today's classrooms. Technology has been a huge expense for schools as well as a big disappointment.

Schools have had a long-standing immunity against the introduction of new technologies. In 1922 Thomas Edison predicted that movies would replace textbooks. In 1945 one forecaster imagined radios as common as blackboards in classrooms. In the 1960s, B. F. Skinner predicted that teaching machines and programmed instruction would double the amount of information students could learn in a given time. Filmstrips and other audiovisual aids were fads thirty years ago, and the television, now seen as a supplier of brain candy, once had a sterling reputation as an education machine.

The public education system has tried repeatedly to extract the potential of the PC. The Congressional Office of Technology Assessment analyzed the evolution of these frustrating attempts. Its report notes that in the early days of the IBM PC, teachers, parents, and school administrators were told that we needed to teach students to program in BASIC, since that tool came with PCs. Then, the focus moved to the computer language Logo: let's teach students to think, not just program. Oops—a few years later, we were told that computers were best used for drill and practice. Then another correction: since PCs are tools, students should be taught word processing. Later phases emphasized curriculum-specific tools, such as a history database or a science simulation, then Web page design, and then the Internet. The progression reads like an implausible story: How can people see the PC's role in education fail and get redefined over and over and over and still maintain the faith? Wouldn't the joke wear thin after a while?

The fortunes of The Learning Company, one of the most successful education software companies, parallel that of education software in general. With popular titles like "Reader Rabbit" and "Carmen Sandiego," TLC was bought by Mattel in 1998 for almost $4 billion. Three years later, it was resold for one percent of that price.

The generous organization that donates a million dollars of PCs to a school district may be killing with kindness. The total cost of ownership of a PC is much more than the cost of the PC itself. The million-dollar donation condemns the school to spending perhaps half that much *each year* forever to satisfy ongoing needs for software, training, support, and upgrades.

The editor of *Issues in Science and Technology*, seeing PCs' educational promise as largely empty, offered these comments about the overemphasis on the digital divide in 2000: "These students who have less access to computers and the Net also have less access to everything else. Why among all their deprivations should we focus on their lack of computers? Is this what separates the underclass from the upwardly mobile? Hardly. . . . At this stage in the development of educational technology, the computer and Net are a condiment or a dessert on the educational menu."

I'm optimistic about the long-term benefit that computers can give to education. However, we should expect more false starts, each with proponents convinced that (despite the failures in the past) they have finally discovered the true educational potential of computers. Expect them to also shrilly proclaim that neglecting the latest approach will dramatically shortchange the future of our children.

Computers in the Home and Office

Computers make it easier to do a lot of things,
but most of the things they make it easier to do
don't need to be done.
—ANDY ROONEY

In the early days of the PC, many articles were written about the uses of this new tool in the home. Perhaps Dad could store his address book there, or Mom could keep her recipes on it. Of course, to access a recipe meant turning on the PC, waiting a minute or more for it to boot up, finding the correct floppy disk, starting the appropriate program, and searching for the desired entry. And what does one do about the fact that the PC is in the office, not the kitchen? The articles earnestly, almost pathetically, wanted to justify the PC's role in the home, but they were premature. The cookbook had been perfected for its job over more than a century and was very good at it.

PCs have improved greatly since then, and they now bring substantial value to many home applications. But even today, just because a PC *can* do something doesn't mean that it's the best tool for the job. A paper calendar, notepad, or address book—or a cookbook—is still tough to beat.

What truly new consumer applications has the PC brought us? Not writing, because the pen and typewriter preceded it. Not the spreadsheet, because the paper namesake preceded it. Not record keeping, because note cards, check ledgers, address books, and pocket calendars preceded it. Of course, computer gaming is without precedent, though here again this is one of the less important areas. Perhaps "editability" is the PC's most important innovation. You can change a single word and reprint the updated document or change a single value and recompute a spreadsheet—certainly an important new ability. Let's give the PC its due, but no more.

In addition to home use, computers have become widespread within the business world. We've seen, however, that doubling a computer's performance doesn't double any useful metric at the user level. Taking this to an extreme, the Productivity Paradox (discussed below) questions whether there has been a net improvement at all.

Think of construction workers a century ago digging with picks and shovels and using horse-drawn wagons to move dirt. How much more productive did they become when given construction equipment such as trucks and bulldozers? Ten times? A hundred times? Consider the productivity boost that textile mills or the factory assembly line gave to the industrial worker. For some it was more than a hundredfold. Now contrast this with the productivity change the computer provided to the average office worker. Can that person get ten times more work done in the same time? Or even two times? Moore's Law is amazing, but it's *only* amazing.

While computers can do impressive things, businesses should be asking whether an electronic presentation gets the job done better than overhead slides, whether e-mail is better than the telephone, and whether a memo with clip art, cute colors, and four fonts is better than the plain text equivalent. Before you respond that it's no trouble to do all these things because the conference room can display electronic presentations, everyone has a computer for e-mail, and fonts and colors are simple to add using your word processor, remember how much all that infrastructure cost. If the worldwide return due to information technology was $3 trillion per year, we'd *just break even*—because that's how much business spends on it!

What fraction of a typical PC's computing power goes to doing office tasks and what fraction to screen savers? How much time is wasted playing solitaire or browsing the Internet? And how much time is spent learning new software, tweaking and customizing, performing maintenance and backing up files, recovering from viruses, helping coworkers with PC problems, wading through unimportant e-mail and spam, and dealing with any of a dozen additional PC support tasks? These costs are tough to pin down, but one survey estimated that over $300 billion worth of time is lost annually to personal Internet surfing in U.S. businesses. Another estimates that the time spent fiddling with computers rather than working on them costs another $100 billion. The PC can be an excellent servant but it is a demanding master.

Concerns with the Productivity Paradox began in the late 1980s when researchers noticed that the escalating costs of computer technology hadn't led to a corresponding increase in worker productivity. Nobel laureate Robert Solow observed: "You see computers everywhere but in the productivity statistics." Corporate computer expenses have gone from negligible to half of business capital investment. As with the school example above, companies don't buy computers, they buy computerization. A computer might cost $1,000. It's a single transaction. But computerization is an ongoing expense that must be paid for year after year.

Industry observers have offered different resolutions to the paradox. Maybe the current way of measuring productivity doesn't capture the benefits of computerization. Maybe we need to give the technology more time and let businesses learn how to best make use of it. On the other hand, maybe much of computer spending is wasted. Although this is a big issue and I propose no resolution here, we should note that a significant net gain (benefits exceeding costs) from computers is anything but obvious and that the progress predicted by Moore's Law—twice as fast every two years—is far removed from any actual benefit that individuals, companies, or the gross national product will realize.

In some ways, the role of the PC has followed the path taken by the washing machine. Washing machines brought about a tremendous increase in productivity. But this increased efficiency played out in a surprising way. Laundry *volume* increased ten times while the time spent on washing stayed roughly the same. In a similar way, word processing has undeniably allowed much cooler documents (to mention just one office application) but has not increased productivity as much as had been hoped.

> *For a list of all the ways*
> *technology has failed to improve the quality of life,*
> *please press three.*
>
> —ALICE KAHN

8 Corrective Lenses

UP TO THIS POINT, WE HAVE LOOKED at the longevity of technological change, explored some of the unexpected ways that technology evolves and affects our lives, examined some of its downsides, and undercut the prevailing high-tech myths. We can now appreciate some of the forces that cause our distorted view of technology: myopia (which causes the birthday-present syndrome) and hype in the press. Now that we know *what* is wrong and *why* we see things that way, we can begin to discover how technological change really works, and we will find a new model to replace the debunked exponential one.

Technology Hierarchy

The generation that had to deal with the greatest changes
in business, commerce, war, and all other aspects of human life
lived in the first half of the nineteenth century,
not the second half of the twentieth.
—STEPHEN AMBROSE, author and historian (1996)

It's only natural to be most interested in today's latest developments—that's where the changes are occurring. As we've seen, however, a look at the relative importance of modern versus older technologies shows that the older technologies were often more important.

As a model for a technology hierarchy, let's review Abraham Maslow's well-known hierarchy of human needs. He organizes needs into levels according to their importance. For example, the need for food is more important than the need for friendship, which is more important than the need to express creativity. Here is one version of the human-needs hierarchy, with the most important needs at the lowest level (level 1):

5. *Self-actualization:* realization of one's potential, creative behavior, acceptance of self and others

4. *Esteem:* self-confidence, independence, prestige

3. *Social:* love, friendships, association with others

2. *Security:* safety, shelter, protection

1. *Physiology:* hunger, sleep, avoidance of pain

In this model, people always focus on their lowest unmet need; only when all needs in a given level are satisfied can the individual move on to address needs at the next level. For example, a contented person working on the top level will suddenly change focus if diagnosed with cancer. Self-actualization is forgotten as the focus turns to more fundamental health concerns. Maslow represented this hierarchy graphically as a pyramid. As we move down, each level is shown wider than the one above it, emphasizing the relative importance of the lower needs.

This hierarchy of needs has a direct parallel with a hierarchy of technologies (see figure 7). As with human needs, not all technologies are equally important. To evaluate the relative importance of two technologies, imagine how life would be different if either were removed. Life would be worse without textiles than without DVDs, worse without steel than without plastic, and worse without electricity than without air conditioning.

Need help in ranking which technologies are more important than others? Spend a night outside without technology—no house, no

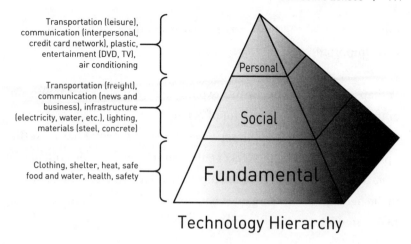

Transportation (leisure),
communication (interpersonal,
credit card network), plastic,
entertainment (DVD, TV),
air conditioning

Transportation (freight),
communication (news and
business), infrastructure
(electricity, water, etc.), lighting,
materials (steel, concrete)

Clothing, shelter, heat, safe
food and water, health, safety

Personal

Social

Fundamental

Technology Hierarchy

Figure 7. Technology Hierarchy. The most fundamental and important needs are on the bottom.

warm furnace, no sleeping bag, not even fleece clothing. When you've mastered that, do it in Minnesota in the winter. Or, for a week get your food from the woods instead of from the store. Shelter and food are much more fundamental than DVDs or the Internet. Refugees cast out of modern homes don't miss television and CDs as much as shelter and access to food and clean water. Worldwide, over one billion people live on a dollar a day, and they have a lot more important things than PCs and Internet access to worry about.

While modern improvements have been made to all levels in the hierarchy, note that the lower needs are addressed by technologies with older roots. Handmade textiles and construction technologies at the lowest level preceded electrical infrastructure and business communication at the middle level, which preceded leisure transportation and entertainment technology at the highest level.

Today, our attention is focused on the upper levels of the pyramid, not because that's most important but because that's where the progress is. We have the luxury of being able to focus on the upper level because technology has already satisfied our lower-level needs. In other words, the top level gets the most attention although it is the *least* important.

Perhaps you still cling to the exponential model. Overcoming an entrenched belief requires a lot of evidence and time. Figure 8 is one attempt to illustrate how biases work. Almost everyone has seen the optical illusion called the Necker cube. Do you see it coming out of the page to the bottom-left or to the top-right? Once you have made sense of the figure by identifying one view, you may find it difficult to switch to the other view.

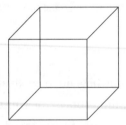

Figure 8. Necker cube

This locking on to a particular viewpoint illustrates the problem we all have in confronting our biases. If you tend to agree with the exponential model, you may be locked into that viewpoint and may be protecting yourself from competing viewpoints. Psychologists call this the confirmation bias: once you have a mental model that explains something, you tend to embrace evidence that supports that model and ignore or reinterpret evidence that doesn't. This is human nature—to challenge everything is to doubt that the next step is reliable or the next breath is safe.

A Zen story provides another kind of illustration. A man wanted to become the student of a Zen master. As they talked, the master prepared the tea. He slowly poured tea into a cup and continued to pour as tea filled up the cup, spilled over, and poured onto the table. And still he kept pouring. "Stop!" the man finally protested. "The cup is full!

It can hold no more tea." The master replied, "And like the tea cup, your *mind* is full. It will accept no new ideas."

Technology's Family Tree

Technology is its own fertilizer.
—DAVID LANCE GOINES,
    artist and writer (1985)

The appeal of the exponential model is that it seems so obvious. With an increasing human population, a steady demand for innovation, and an increasing store of knowledge and technologies, one would expect that there would be more fronts of research than ever and technology would be making an exponentially increasing impact on the average citizen. We have positive feedback: the more products we invent, the more synergy is possible and therefore . . . the more products we invent! This model of technological innovation would end up looking a little like a family tree (see figure 9): a single couple creates a few children, they create many grandchildren, and so on.

Though it may seem logical, this ever-expanding family-tree model isn't the way things work. Rarely do we see a completely new, unprecedented technology spring forth out of nothing. More often, new technology is an improvement over the old—it does the same thing, just better. It doesn't form an additional branch on the family tree; it *replaces* an old branch. Once superseded, that old technology quickly loses relevance. For example, there is little value today in cast iron or stoneworking techniques, in barbed wire or railroad brake patents, or in ax or telegraph technology. From waterwheels, sailing ships, and stone masonry to paper clips, arc lights, and butter churns, history shows us technologies that were leading edge in their day but have since become outmoded. We see the baton passed from one technology to its successor, with the old technology largely discarded: from LP to compact disc, from steam locomotive to diesel locomotive, from piston engine to jet engine, from manual labor to machine production. The result is not a rapidly

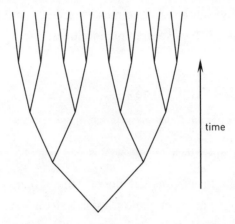

Figure 9. Simple but incorrect family-tree
model of technological innovation

increasing pile of technology to work with but a collection of the state of
the art in each technology that is fairly constant in size. The pile that is
truly growing is that of the outdated and discarded technologies.

Another problem with this simplistic family-tree model of technology's
evolution is that it assumes that all innovations (new branches in the
tree) are equally important. This ignores the Technology Hierarchy,
which shows that the most basic and important technologies were
developed long ago. We see the same phenomenon in many other
fields. For example, in the 540 million years since the beginning of
biology's Cambrian Explosion, the fundamental forms of animal life
(the twenty or so phyla) developed in roughly the first 1 percent of that
time period, and the evolution during the remaining 99 percent were
all variations on those themes. Similarly, many literary critics have
suggested that the basic stories have long since been invented and that
literature, theater, and film simply reinterpret these perennial themes.
The basic elements of architecture (the cable, beam, arch, and so on)
were all invented thousands of years ago, as were the simple machines
of mechanics, such as the lever, screw, and pulley.

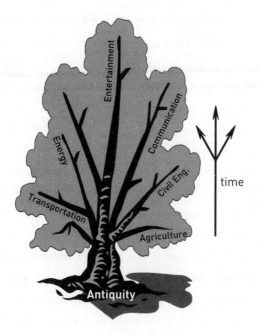

Figure 10. Real-tree model of innovation. The technology we use today is at the tips of the branches. Dead ends and obsolete technology that are forgotten and no longer play a role are the dead branches.

New forms of a technology sometimes must wait for supporting technologies to catch up, but the delay is rarely due to a lack of imagination. Usually, all niches are quickly searched, in Cambrian Explosion fashion. The internal combustion engine was invented and soon used in the car, truck, and airplane. Soon after the introduction of the radio, many variations were tried, including early versions of portable radios, car radios, and boom boxes. The basic ways of paying for broadcasting were soon explored: advertising, customer pays (as used by the BBC), and government funding plus public donations (PBS). We find similar creative explosions in the early years of the printing press, the electric motor, and other inventions.

The Technology Hierarchy allows us to correct the family-tree model. Imagine a real tree, old and weathered, instead of the idealized family tree (see figure 10). As with the simplistic family tree, time goes from antiquity at the base to the present at the leaves. The main branches of our technology tree are all very old (transportation, entertainment, energy, manufacturing, construction, and so on) and have forked from the trunk near the base. Technology's Cambrian Explosion occurred thousands of years ago, when the fundamental forms of these main branches were established.

When a branch forks into two branches, one fork usually dies. The old fork can die when a new fork supersedes it (horse by car, telegraph by telephone, steam engine by diesel, propeller by jet, LP by compact disc, and so on). An entire branch rarely dies out—if something was worth doing in the past, we just find better ways of doing it. Instead, an old technology yields to its successor. Another possible result is that the *new* fork dies when it represents a fad or experiment that didn't catch on: the gramophone as a dictation device, the telephone for broadcasting news or music, concrete to build houses, nuclear weapons for construction, 3D movies, movies to replace books, and so on.

Take consumer audio recordings. Cylinders gave way to 78s, which gave way to LPs, 45s, and reel-to-reel tape, which gave way to CDs and cassette tapes. Now tapes are fading, with digital (MP3) players replacing them, and digital players may replace CDs as well. Along the way are short branches to extinct experiments such as Digital Audio Tape (DAT), quadraphonic sound, 8-track tapes, and Digital Compact Cassette (DCC). We have gone from one active technology (cylinders) to two (CDs and MP3)—not a big jump. Compare that with the large number of extinct forks on the branch. Once cylinders, 8-track, and DAT were dead, they stayed dead. Innovation happens just at the tips of the branches and the trash pile of old technologies never becomes a major source of ideas. While we do occasionally see new branches (consumer audio, radio, and aviation, for example), most of the innovation is replacement and not something fundamentally new, and the number of active branches today is little more than it was a century ago.

Let's contrast two extremes of tree models. First, consider the simple family tree, which doubles the number of branches with every generation. Next, imagine the tree of all life forms, from which it is estimated that 99.9 percent of all species that ever lived are now extinct. The first tree is extremely dense, the second wiry and sparse. The technology tree is somewhere in the middle, growing a bit as new technologies (rarely) enable something completely new, but in general much sparser than the stereotypical family tree.

Technology Spotlight

If man had never charted a path into the unknown,
he would never have discovered the known.
—MIRIAM R. G. GOULD

With the exponential curve model discredited, we need a new model to explain the nature and rate of technological change. I propose the Technology Spotlight. To see how this model would work, imagine a diorama that contains all technologies. In this model we see workers building a house (representing construction), a blacksmith (metallurgy), a team of horses pulling a wagon (transportation), and so on. The diorama has a slider to let you choose any time over the past thousand years. As you move the slider to push the exhibit through time, one or more spotlights illuminate different areas that are undergoing extraordinary change. In the 1200s, the spotlight is on construction as cathedrals are pushed to new heights. After about fifty years, cathedrals are still being built, but the revolutionary phase has passed and so the spotlight fades. In the late 1400s, the spotlight shines on communication as the printing press sweeps Europe. By 1500, that spotlight fades. Of course, beyond this date the press still existed and improvements were being made, but the spotlight highlights social *change*, and the period of abrupt change was over. At any time, one or a few of the dozen or so major technologies are illuminated and improving quickly, while the others that remain in the dark are improving very little.

The spotlight can fade without warning and may return later. For example, we see the spotlight shining on construction during the innovative phase of cathedral building, moving away, and then returning between roughly 1880 and 1930, during the skyscraper's revolutionary period. It shone on communication in the early days of the printing press and then faded as that technology matured. When high-volume steam presses were introduced in the early 1800s, newspaper volume exploded. The spotlight had returned, and printing technology was again changing society.

The spotlight can shine on several technologies at once. The early 1800s saw the spotlight on communication (printing, and later the telegraph) concurrent with transportation (railroad). In the early 1900s, it was on transportation (airplane) along with construction (skyscrapers, dams, and bridges). Today, it's on communication (Internet and cellular communication) plus computation.

One sure sign that the spotlight has moved on is the labeling of a technology as "mature." Stock analysts and economists divide industries into categories such as chemicals, energy, manufacturing, metals and mining, transportation, utilities, and so on. Though not called high-tech—at least not anymore—these are still technology-intensive industries. They *were* high-tech when the spotlight was on them in their day.

Of course, all of the post-spotlight technologies are still in use (with a few exceptions, like the telegraph), and much money is still being spent on research to improve each one. However, the social *change* that each causes has slowed. When a technology is popping up in new and surprising applications, being discussed in the media, and causing delight and frustration in people's lives, it's in the spotlight. When that change slows, when the technology becomes part of the environment, and when it is only noticed in its absence (a power failure or a road closed for construction, for example), the spotlight fades. That technology is largely ignored even though it may still be useful and improving—even fundamental. The spotlight is a metaphor for the conditions that make innovations happen—it's society's technology muse.

Let's imagine a situation in which there is no pressure of technology change. Suppose technological change had stopped twenty years ago. That is, for the past twenty years, not a single technological improvement had been invented, patented, carried out, or discussed in the press. Existing products would have traversed their growth curves and all consumers wanting products could have bought them. All spotlights would therefore have dimmed out. With technology well assimilated into society and product penetration stable, stress from technology change would be negligible. The amount of technology would be the most ever, but the number of areas under the spotlight, being proportional to the amount of *change,* would be zero.

Note also that the spotlight highlights *social* change. Putting humans on the moon was amazing and took a huge amount of innovation, but it did not change people's lives. By contrast, the Web has had much more effect on our daily lives but required much less innovation.

The spotlight model encompasses the exponential model. It acknowledges that individual technologies can grow exponentially (for a while, anyway) but accepts the inevitable slowdowns as government policy, economic events, or just tough technical problems hinder consistent progress.

Today's technology is built on a foundation that was laid by the technologies of the past. Imagine being carried 95 percent of the way to a mountain summit, climbing the last bit alone, and then bragging afterward about your solo effort. True, when at the top you were the highest, but keep in mind the small fraction of the total effort you contributed. Similarly, we too often focus on our own modest contributions to society's technology stockpile, ignoring the legacy we inherited. Even the great Isaac Newton appreciated the work of the pioneers who came before him. He said, "If I have seen further [than others] it is by standing upon the shoulders of Giants." We, too, need a similar appreciation of the foundation on which modern society is built. Today's technology is truly standing on the shoulders of giants.

The press, the machine, the railway, the telegraph are premises
whose thousand-year conclusion no one has yet dared to draw.
—FRIEDRICH NIETZSCHE,
The Wanderer and His Shadow (1880)

PART II. THE MORE THINGS CHANGE . . .

9 For Better or For Worse

"MY NAME IS OZYMANDIAS, king of kings: Look on my works, ye mighty, and despair!" This quote from Percy Shelley's poem "Ozymandias" (1817) is the inscription on a statue of a ruler better known today as Ramses II of Egypt. The once-mighty statue now lies broken and scattered in the desert. The poem ends: "Round the decay of that colossal wreck, boundless and bare, the lone and level sands stretch far away."

Like the works of King Ramses, something that seems a permanent feature can fade away, ignored and forgotten as it is replaced by something new. Technological wonders that are new and exciting today are taken for granted tomorrow. In a similar way, our interaction with technology can be like a mediocre marriage, where initial infatuation fades and we take our partner's good points for granted—for better or for worse.

In part I we surveyed the technology landscape, finding and fixing important misconceptions. In part II we will focus on the constancy of technological change. Like each generation's conceit that it invented sex, too often we deceive ourselves that we're the only generation struggling with and delighting in technology change. Rather, this survey will show a very long and interesting history.

Ages Through the Ages

Nuclear powered vacuum cleaners
will probably be ready within ten years.
—founder of Lewyt Corp.,
  vacuum cleaner manufacturer (1955)

As we survey the impressive landscape of our Information Age, we should have a sense of déjà vu. Remember the Atomic Age, when nuclear power promised to change everything? Nuclear power had been successfully applied in submarines, ships, and power plants in the 1950s. From this success, experts predicted nuclear-powered airplanes and cars and huge nuclear-powered oil-tanker submarines. David Sarnoff, chairman of RCA, predicted "atomic batteries" before 1980. *Business Week* cautioned in 1947: "Today no prudent businessman, no prudent engineer dares make plans or decisions reaching more than about five years into the future without at least weighing the possibility that the basis of his planning may be upset by the commercialization of discoveries about the atom." (This sounds like similar overenthusiasm about the Internet.)

In 1954, the chair of the Atomic Energy Commission predicted "unlimited power . . . in fifteen short years. It is not too much to expect that our children will enjoy in their homes electrical energy too cheap to meter, will know of great periodic regional famines in the world only as matters of history, will travel effortlessly over the seas and under them and through the air with a minimum of danger and at great speeds, and will experience a life span far longer than ours. . . . This is the forecast for an age of peace."

Project Plowshare, inspired by the biblical quote "they will beat their swords into plowshares," proposed to apply not just nuclear power, but actual atomic bombs to earthmoving projects such as reshaping harbors, digging canals, or mining. There was also a curious fixation with weather control: predictions that nuclear power would eliminate hurricanes, tornadoes, and droughts; heat cold regions of the world; prevent rain on the weekends; and ensure only sunny weather for sporting events.

The public lost interest in atomic energy with the success of the space program in the 1960s. We had entered the Space Age! It would not be long, we were told, before we would be exploring Mars and vacationing in moon bases. Again, the enthusiasm soon faded. The end of the Skylab space station program illustrates the abruptness of this change. Less than five years after the first moon landing, the Skylab 3 crew returned to Earth, untelevised and little noticed. Jeff MacNelly's editorial cartoon for that day showed a bewildered astronaut looking out the open door of a space capsule bobbing in the ocean with darkness falling. The caption read, "But *somebody* must have heard we were splashing down. . . . Try Houston again." Skylab was allowed to fall back to Earth and burn up five years later.

Now, with the widespread use of the PC, we are in the Information Age. The nuclear genie has not been put back in its bottle and the knowledge needed to land on the moon has not been forgotten—so what happened to the previous Ages? The Space Age, the Atomic Age, and previous "Ages" that society has labeled for fundamental new developments like jets, cars, electricity, or railroads faded away for various reasons, some unique to each period. Rather than spend too much time on the *why*, let us learn history's lesson: that each Age is impermanent, making its impression for only a decade or two before it fades in importance. Labeling a period as a Technology Age is one way society identifies the Technology Spotlight.

The business world is not immune to fads, either. By the 1980s, Japanese companies had made tremendous inroads into shipbuilding, steel, cars, and electronics. Searching for a response, U.S. industries fixed on the fanatical devotion to quality in the typical Japanese manufacturing company. Extremely high quality became the perceived solution to most U.S. marketplace problems.

In the '90s, the knee-jerk response switched from quality to Internet Time. There was nothing wrong with high quality, of course, but we had to get that development cycle time down. The Internet was here, and products that took years to get to market were now apparently taking months.

Like quality, computerization too has become unremarkable. Most manual tasks that can be profitably computerized have been. It isn't a bragging point when everyone does it. Similarly, the idea of "putting the company on the Internet" is no longer revolutionary or even noteworthy because it is so commonplace. Remember when appending ".com" to company names—not just those of Internet companies— was a short-lived fad in the late 1990s and good insurance for a successful public stock offering.

Computers, the Internet, and everything else associated with the Information Age are important, of course, but remember the short lives of previous "Ages." Information is in the spotlight now, but it won't be there much longer. Perhaps within a decade we will have a new issue du jour as we enter the Age of Robotics, Genetic Engineering, Quantum Computing, or Artificial Intelligence. Or, maybe it will be Nanotechnology, Virtual Reality, or Superconductivity. Or Solar Power, Fusion Power, or Hydrogen Power. With the arrival of this new Age, whatever it is, "Information" will fade into the background as an essential but unremarkable component of society, like electricity or antibiotics, cars or concrete.

Technology Defines Society

Have I done the world good,
or have I added a menace?
—Guglielmo Marconi (1874–1937),
  wondering about the impact of his
  invention, the radio.

In 1987, North Korea began constructing what was planned as the world's tallest hotel. After spending several years and close to a billion dollars, the exterior of the 105-story Ryugyong Hotel was finished. Unfortunately, no funds remained to complete the interior, and the building has never opened. It sits forlorn and empty—or as a monument to the glory of Marxist-Leninist-style proletariat internationalism, depending on your perspective.

Bragging rights from having a technology marvel are sought by societies worldwide. The North Korean government would have delighted in having its country known as the home of the world's tallest hotel. Malaysia is proud of its Petronas Towers, once the tallest buildings in the world. Chicago was quick to point out that its Sears Tower still had the tallest occupied floor, although Taiwan has since taken that title. The Soviet space program was a morale booster in the USSR. By the same token, what is a source of pride to you becomes a target to an enemy: the World Trade Center and the Pentagon were chosen as terrorist targets on September 11, 2001, because they were big, visible, and symbolic. Though gone, those towers are vastly bigger symbols than they ever were when standing.

Think of the Empire State Building or the Statue of Liberty and you think of New York; think of the Eiffel Tower or Notre Dame and you think of Paris. There remain plenty of seemingly impossible projects to aim for—a sea-level Panama canal without locks or a mile-high building, for example. Tough projects like these can challenge a society and, once achieved, give pride and prestige to its people. Using the space program as a way to define the country, President Kennedy said in 1962, "We choose to go to the moon in this decade and do the other things, not because they are easy, but because they are hard."

Big projects have been sources of pride and have helped define societies for millennia. The best known of these society-defining projects may be the Seven Wonders of the Ancient World. This list was compiled in the second century BCE and includes the primary architectural marvels of the time from the eastern Mediterranean, including the Pyramids of Giza, the Colossus of Rhodes, and the Hanging Gardens of Babylon. But the world is full of engineering wonders. Stonehenge and the statues of Easter Island, Angkor Wat and the Roman Coliseum, the Great Wall of China and the Sphinx: these and others also helped define and enhance the reputations of their own societies.

Technology and Popular Culture

Whoever wishes to foresee the future must consult the past;
for human events ever resemble those of preceding times.
—NICCOLÒ MACHIAVELLI, *The Prince* (1513)

Mary Shelley's 1916 novel *Frankenstein* tells the story of a scientist who brings the dead back to life. Its subtitle is *The Modern Prometheus,* likening Victor Frankenstein to the god who dared to bring the gift of fire to humanity. The scientist's hoped-for gift goes very wrong, however, and the novel explores the public's concerns in Shelley's time over the impact of science and industrialization on society.

Literature has reflected on and explored the concerns of the day for centuries. Just a few years after the American Civil War, during which civilian technologies were drafted into war service, Jules Verne wrote *20,000 Leagues Under the Sea* (1869). In it, Captain Nemo tries to recapture a few of the demons released from Pandora's Box by building a submarine to destroy warships. *The Time Machine* (1895) by H. G. Wells considered the evolution of mankind at a time when Darwin's theory of evolution was taking hold. Sir Arthur Conan Doyle's *The Lost World* (1912) told of the discovery of a lost dinosaur habitat in an isolated part of South America, at a time when archeologists were uncovering the fossils of enormous animals and explorers were uncovering new parts of the earth. Compare this with the approach to dinosaurs in *Jurassic Park* (1990). At that time, new territories were no longer being discovered, but genetic research was exploring new domains, and this played a central role in the book.

After earlier books about the individual mishandling of technology, such as *Frankenstein* and *The Strange Case of Dr. Jekyll and Mr. Hyde* (1886) or early science fiction such as *The Time Machine* and *A Journey to the Center of the Earth* (1871), literature and culture moved on to explore the larger issue of technology as a detriment to society. Movies such as *Metropolis* (1927) and *Modern Times* (1936) showed the darker view of a society with technology out of control. *Brave New World* (1932), *Nineteen Eighty-Four* (1949), and *Fahrenheit 451* (1953) illustrated other negative utopias in which varying amounts of technology were used to keep people in their places.

Modern literature such as *Future Shock* (1970), the many books predicting "the coming crash" or "the coming end" of something or other, and movies such as *The Terminator* (1984) or *The Matrix* (1999) are recent examples of more contemporary technology anxiety. Literature changes to reflect what's on the public's mind and helps preserve a record—a long one—of how society frets about technology. Hype can convince us that not only is technology moving so fast that it's leaving us all in the dust, but that it's also on the fast track to hell.

Popular culture is another area that reflects our priorities. Marvel Comics launched the *Spider-Man* comic in 1962. There we learned that Peter Parker became Spider-Man after being bitten by a radioactive spider. Forty years later, the movie version also shows Peter obtaining his superpowers from a spider bite, but this time it is a *genetically modified* spider. Hollywood tracks what's on the public mind: radioactivity wasn't interesting anymore, but the dangers of genetically modified animals were. *The Hulk* comic (also 1962) shows Bruce Banner transformed into the Hulk by radiation from a nuclear explosion, but in the movie version (2003), radiation triggers *nanobots* that make the transformation. We see another cultural fossil of the public mindset when in the 1967 movie *The Graduate* the title character is given hushed, almost clandestine career advice: "Plastics."

Western culture can become more sinister when, through technology, it is conveyed to another culture. Television came to the Basque region in the Pyrenees Mountains in southwestern France in the late 1960s.

Euskara, the ancient Basque language that survived invasion by the Romans and by the Celts before them, has had difficulty competing with modern French culture that effortlessly penetrates the remote mountains on TV signals.

Bhutan, the tiny land hidden high in the Himalayas and ruled by its Dragon King, has had roads, electricity, and public schools for only a few decades. In 1999, it became the last country on earth to get television. This gentle country with gross national happiness as its guiding principle has had a difficult time coping with televised violence and the lure of western goods.

Popular culture has tried to predict the changes that we can expect in our own culture, though with limited success. *The Jetsons* TV cartoon, shown first in 1962 and set a century in the future, is a familiar bit of contemporary science fiction. George Jetson works three hours per day pushing buttons as a "digital index operator." Calisthenics for Jane Jetson are morning button-pushing exercises. These extrapolations of the 1960s' predictions about the diminishing amount and intensity of labor may yet come true, but that seems unlikely anytime soon. The number one song for 1969, "In the Year 2525," predicts atrophy of the human body, actions controlled by pills, and robots replacing humans. ("In the year 5555 / Your arms are hanging limp at your sides / Your legs got nothing to do / Some machine's doing that for you.") The novel *2001* (1968) was also off the mark in predicting human-like intelligence for the computer HAL. Even its ominous (though perhaps accidental) warning in the name HAL—take the succeeding letters of the acronym HAL to get IBM—seems quaint now.

Technology Can Be a Tight Fit

Science explores: Technology executes: Man conforms.
—motto of 1933 World's Fair in Chicago

Around 1970, I saw a movie short in which a clerk fills in a form with a customer's personal information. The clerk first asks for the customer's name, street address, and city. Then the clerk asks for the apartment number, zip code, and phone number. Then come requests for more numbers: years of employment, years of schooling, social security number, age, income, height. The dialog proceeds until it finally degrades into a meaningless exchange of numbers, with the participants communicating like computers. A self-portrait by Paula Scher makes a similar statement. In the outline of a head are the dozens of numbers that identify her—birth date, passport number, driver's license, car license, marriage license, insurance and credit card numbers, frequent flyer numbers, prescription numbers, and on and on.

It may be difficult to remember the concerns raised during the 1960s and '70s, when mainframe computers were first being used in business. At that time, computers seemed ominous, and people didn't want to be considered "just a number." Social Security numbers were becoming identification numbers for lots of new purposes, and the computer punch card became a metaphor for the transformation of an individual into data. To be more easily processed by computers, products were branded with bar codes. Some saw IBM (or at least their machines) as the personification of Big Brother.

Earlier decades felt the imperfect fit of technology, just as we do. The praise given to and concern expressed about the early telephone by social commentators of the time might sound familiar today if the word *Internet* replaced *telephone*. Some saw the telephone linking faraway family members and strengthening society, but others feared a shallower community, with personal encounters replaced by telephone calls. Perhaps people would gravitate to others like themselves and shun the wider community; perhaps relations would become more impersonal and superficial. In 1899, one Englishman stated that anyone able to telephone anyone else was to be feared "by the sane and sensible citizen."

Quick access to worldwide news also raised issues during that time that sound like concerns about technology raised today. Alarmists worried that journalism brought too wide a spectrum of information to the reader compared to the norms of a century earlier—or even a generation earlier. Was it desirable to know within a day every significant event worldwide? Was news becoming too homogenized? Was the stress ("unnatural excitement" in the words of the day) imposed on the business worker by rapid news and rigid time demands too much to bear? Max Nordau, a widely read social critic of the late 1800s, predicted that the new demands imposed by rapid communication would take a century for people to adapt to.

Sleep might seem to be a natural refuge that technology cannot reach, and yet we find significant differences between the twentieth century's technology-enhanced nights and those of the past. From antiquity through most of the nineteenth century, there wasn't much to do at night except perform the limited activities enabled by a lamp or candle, or to sleep. People typically slept nine hours a night and night-shift work was rare. A recent article in *Smithsonian* observes that artificial lights and the demands of work disrupt what may be a more natural sleep pattern. Unburdened with light, people often slept for about four hours and then awoke. They might think, talk, pray, or even visit for an hour or more, then sleep for the remainder of the night. There were even accepted terms for this: "first sleep" and "second sleep." With alarm clocks and electric lights in use since the late 1800s, natural sleep patterns are often interrupted, and American adults now sleep only about seven hours per night. Workers' convenience has become less important than maximizing capital investment, and some factories and stores are staffed twenty-four hours per day.

Today we worry about the downsides of security cameras in public places or location-tracking ability in cell phones, but concern about privacy is also not new. Around 1900, citizens fretted about microphones being used to secretly record conversations, cameras taking unwanted pictures, electric doorbells and the telephone allowing strangers easy access into the home, and radio messages being more vulnerable to eavesdropping than those sent by telegraph.

By now we have reached a standoff with these technologies: their privacy downsides either are addressed with laws or conventions, or they are simply understood and accepted by the public. Today, recent technology has created new concerns such as online privacy, identification theft, and location tracking. Whether these issues are resolved quickly or get worse before they get better, it is likely that they will eventually fade in importance. Consider credit cards as a precedent. In the 1970s, important issues were being hammered out in this new consumer product. How much interest can be charged? Who's responsible for fraudulent charges to my account? What if my card is stolen but I don't report it? What if my credit report contains errors? Banks fought the consumer-friendly legislation, which burdened them with most of the risk, but the result is a system with strong consumer confidence and through which hundreds of billions of dollars are spent annually. This combination of legislated safeguards and the public's growing familiarity is common to the success of new technologies that initially worry us.

Social Stereotypes

[Television] is an art which shines
like a torch of hope in the troubled world.
It is a creative force which we must learn
to utilize for the benefit of all mankind.
—DAVID SARNOFF, at the public introduction
of RCA's television in 1939

I wish goddamned television had never been invented.
—EDWARD R. MURROW, radio journalist

Irene and Vernon Castle were popular dancers in the decade from 1910 to 1920, usually playing to sold-out houses. Until this time, well-dressed men typically carried a pocket watch, while women often wore a "bracelet watch" (wristwatch). When Vernon Castle switched to a wristwatch, many contemporary men did the same. With his widespread popularity he was able to eliminate the stigma attached to what had been a feminine accessory. Before any new product can be widely adopted, it must overcome any prejudices against it.

What does technology say about its user? When I went to high school, wearing a calculator on your belt meant "I'm a nerd" as much as thick glasses and high-water pants. Before that, it was a slide rule. In its early days, a pager meant "I'm a doctor," but in some circles it later came to mean "I'm a drug dealer." Cell phones are cool when seen as exclusive but not so cool if seen as an intrusive around-the-clock leash held by your company. Before sunscreen was invented around 1940, only someone obliged to work in the sun had tanned skin. Sunscreen allowed beachgoers to tan more safely, and the statement made by bronzed skin went from "I'm poor" to "I'm healthy."

Try to remember the first time you saw someone walking in public while talking on a cell phone. It's commonplace now, but I remember wondering if the person was talking to himself. More recently, earpiece microphones allow people to talk without holding the phone up to their ears, making them look even more like they are conversing with the voices in their own heads. Maybe you remember feeling self-conscious as the phone user in these situations if people stared at you.

Just because a product can be built doesn't mean that it ought to be. The social obstacles to a new product can be more overwhelming than the technical ones, and products often fail because they are out of step with consumers' social needs. This was the case with the videophone, and moving sidewalks don't sell when people are concerned about fitness. Sales of Citizen Band radios collapsed in 1977, a few years after they boomed, because the CB didn't address any real need. Videotex experiments also failed. Other missteps have been disposable paper clothes, 3D movies, nuclear bombs for earthmoving projects, and concrete for homes.

Status (good or bad) can come from the clothes you wear, the car you drive, or the technology you carry. A Cadillac sedan makes a different statement than an old VW bug, just like taking notes on the latest laptop makes a different statement than taking notes on paper.

Social Conventions

When asked, "Do you want to stop progress?"
technology critic Howard Rheingold replies, "Progress toward what?"

An enterprising William Dockwra implemented a revolutionary new mail system. It included hundreds of mail delivery stations from which mail was picked up hourly. Deliveries were made anywhere in the great city at least four times per day; they were made ten times per day or more in the busiest sections. The charge was one penny per pound of weight (or twice that rate for deliveries up to ten miles out of town), and each letter or package was insured. Where and when was this postal utopia? It was London in 1680.

Other cities in Britain adopted the penny post, but for mail sent out of the city, the rate increased with distance and the recipient paid the postage. It took another innovator—Rowland Hill is usually credited with this—to realize that the system's primary cost was not transportation but the staff required to weigh incoming mail and collect postage, and he proposed adopting the efficient penny post system nationwide. Adhesive stamps were already in use to document other payments, and the famous Penny Black stamp of 1840 inaugurated the first modern nationwide mail system.

International mail delivery was standardized a few decades later. One of its rules is that stamps must bear the name of the issuing country—though an exemption was granted for Britain, since it had created the innovation.

We see a rough parallel today with the Web's top-level domains. An international standard assigns a two-letter code to every country. It's .fr for France, .ru for Russia, and .us for the United States. However, U.S. sites are much more likely to use .com, .org, .gov, or .net. The Web became international only after use of the more common domains became established in the United States. In a repeat of the early days of international mail, the minimal use of .us sites acknowledges the Web's early prominence in the United States.

Customs like this are negotiated daily as new ideas migrate into habits. For example, do we write *e-mail* or *email? Information Superhighway* reverted to *Internet,* but has this now evolved to *internet?* When conveying a Web address, at first the entire address was carefully spelled out, letter by letter: "http://www.acme.com." Now, a simple "acme.com" is not just acceptable but preferred. We debate the ethics of sharing copyrighted material or the appropriateness of using a cell phone where strangers can overhear. Some writers use e-mail emoticons such as ";-)" and acronyms like IMO ("in my opinion") to soften messages and avoid misinterpretations. In chat rooms, we tolerate nicknames instead of real names, which provide the same anonymity that handles did for CB radio users in the 1970s. "Newbies" to various parts of the Internet might be gently steered to the FAQ (list of Frequently Asked Questions), or they might be curtly told to RTFM (Read The F***ing Manual).

With new technologies, everyone's a newbie and new customs must emerge from the old. For example, when you use a videoconferencing system, how important is your appearance? Is it more formal than a phone call or a face-to-face meeting? How do you start when talking with a stranger for the first time? Who hangs up first?

At least the videophone has a precedent. The telephone was a bigger leap with fewer precedents to draw on. It often came with a manual, users were discouraged from chatting, some were paralyzed with stage fright, and even the greeting was debated. Emily Post advised that instead of answering a phone with a polite "This is John Doe," the anonymous "Hello" was safer, and parents were cautioned that with this new tool, strangers could bypass locked doors to reach their children. A joke of the time had a country boy responding to a caller's increasingly agitated "Are you there?" by nodding his head. The bumpkin had a hard time adapting to this network that projected one's virtual presence to a distant place—a network that preceded the Internet by over a century.

Peter Drucker observed this about an 1882 telephone conference convened by the German post office: "The topic—and only chief executive officers were invited—was how not to be afraid of the telephone. Nobody showed up. The invitees were insulted. The idea that they should use telephones was unthinkable. The telephone was for underlings." Of course, the telephone has long since moved into common use—society has changed its definition of the telephone from something used by secretaries and receptionists to something so essential and simple that everyone uses it.

This same transition occurred more recently for the typewriter keyboard. Just a few decades ago, the keyboard was the domain of secretaries, but today most people are comfortable with keyboards and even executives write much of their correspondence themselves. Typewriters were hardly as indispensable when first introduced, and the typewriters of the 1870s were at odds with the social conventions of the day. A typewritten letter might well have been more legible and faster to produce, but it seemed cold and impersonal at a time when people expected handwritten correspondence. Some recipients discarded them, thinking them junk mail, while others were offended, concluding that the company assumed they couldn't read script and felt obliged to typeset the letter.

Curious things can happen when new technology rubs against existing customs. In the early days of the radio, BBC announcers were required to wear black tie when reading the six o'clock news. The Japanese speaker may feel obliged to bow when using the telephone even though the recipient isn't able to appreciate the gesture, and many people use hand gestures on the phone as though they were speaking in person.

Etiquette has even been a factor in warfare technologies. Early machine guns were used in the Civil War, but even in World War I they were not used to their fullest capacity. Traditionalists insisted that machines must not preclude opportunities for individual heroism or the glorious cavalry charge. Not all changes in warfare have been toward greater carnage. Limits on the use of poison gas were agreed to after the excesses of World War I, and treaties between the United States and the Soviet Union curbed the supply of nuclear warheads. During the Revolutionary War, we see a clash between British customs developed on big European battlefields and the American techniques of camouflage and firing from behind trees. The British felt that this was not only ungentlemanly but cowardly. Earlier still, military life was guided by strict codes such as medieval chivalry and the roughly contemporaneous Japanese code of Bushido.

Where technology pushes too far, society pushes back. In a tangible rejection of technology hype, several dozen Italian "Slow Cities" have rejected many of the consequences of modern technology to preserve important elements of their centuries-old lifestyles. Banned within these cities are twenty-four-hour supermarkets, Internet cafes, neon signs, advertising posters, TV antennas, car alarms, and (of course) fast food, plus other elements of technology seen as more harmful than beneficial. Where technology oversteps its bounds, push back!

If there is technological advance without social advance,
there is, almost automatically, an increase in human misery.
—MICHAEL HARRINGTON, social scientist (1962)

10 Playing with Matches

EXTRAORDINARY RAINS HIT HENAN PROVINCE in central China in August 1975. Dams built to handle a flood expected no more than once every five hundred years collapsed, increasing the load on downstream dams. In all, sixty-two dams failed. A flood several miles wide and racing at thirty miles per hour surged out of its river valley and across the plains, killing 85,000 people. Another 100,000 died in the aftermath due to unsafe water and famine, and a total of eleven million people were affected.

But this pales compared to China's 1931 Huang He (Yellow River) disaster. This river frequently floods, and levees have for centuries tried to keep it within its banks. Because it gradually fills its riverbed with the silt that it carries, the levees must be frequently raised to keep it under control. Eventually, the river bottom can be higher than the surrounding countryside. In 1931 the river broke through; between one and four million people died, the deadliest natural disaster in history.

We sometimes find ourselves in a technology cage of our own making. Although we're dependent on technology that can be unhealthy or otherwise dangerous, there is hope. After unfortunate incidents due to shortsighted, unsound practices, we sometimes see the light and change our ways.

Environment

Tug on anything . . .
and you'll find it connected to everything else.
—JOHN MUIR, naturalist and writer

The Cuyahoga River enters Lake Erie at Cleveland, Ohio. One summer day in 1969, a fuel spill on the Cuyahoga caught fire. The startling paradox of a burning river caught America's attention and became a nationwide story, and the Clean Water Act passed three years later. Cleveland, now with a substantially cleaner river running through it, would prefer to forget this bit of its history, but the 1969 fire remains a monument to environmental irresponsibility.

Ironically, the river was already improving at the time and had experienced much worse fires in the past. In the early days of oil refining, when kerosene was widely used for stoves and lamps, the gasoline component had no value and refiners dumped it into the nearest convenient waterway. Significant fires on the Cuyahoga and other nearby rivers date back to the 1860s.

Technology has had a broad and long-term impact on the environment. Waterways have been dumping grounds for thousands of years, and only recently have we become aware that nature has limits as a receptacle for pollutants. Air pollution has also been a consequence of technology, sometimes a fatal one. London experienced a cold snap in early December 1952, and more coal than usual was burned to keep homes warm. A temperature inversion and stagnant air soon turned London's traditional fog into a killer smog. During the day pedestrians in some parts of London couldn't see street signs or even their own feet, and many lost their way. Some drivers abandoned their cars and walked. Indoor concerts were canceled because audiences couldn't see the stage. Death-toll estimates of the four-day Great Smog of 1952 start at four thousand. This tragedy helped bring about the Clean Air Act of 1956, an important step toward London's much-improved air quality.

Lesser versions of the Great Smog date back to a century before, but the problem of smog in London has an even longer history. Public complaints about air pollution date back to the 1200s, and one writer wrote in 1661 of the "Hellish and dismall cloud" of coal smoke that lay over London. Coal was the fuel of the Industrial Revolution and consumption increased a hundredfold from 1800 to 1900, dramatically worsening air pollution.

The increased use of coal was a direct result of another environmental problem, deforestation. Beginning in medieval times, deforestation had begun to change the European landscape. Viewers of the Tour de France may remember Mont Ventoux (Windy Mountain), a tough bicycle climb to a bare and rocky summit. The summit was once covered with forest, cut down long ago to build ships for the French navy, leaving a permanent scar. In Britain, once the larger forests were gone and firewood was scarce, local coal (of poor quality) became the next best fuel source.

Heavy metals are another source of pollution. As awareness of this problem increased in the 1960s, laws were enacted that improved the situation. Lead is no longer allowed in paint or gasoline, and cadmium is being phased out of rechargeable batteries.

Heavy metal pollution used to be an even worse problem than it is today. The Mad Hatter from *Alice in Wonderland* caricatured real hat makers of the 1800s. To make beaver hats, the fur from beaver skins was removed in a process that used mercuric oxide, and hatters often developed mercury poisoning. The illnesses of some painters, like Goya and Van Gogh, may have been caused by the lead or mercury in their paints. During Isaac Newton's time, *tasting* new compounds was considered a useful form of analysis, and his many alchemical experiments exposed him to mercury in ways now considered very dangerous. The cruel rages of Russia's Ivan the Terrible may have been caused by ill-advised mercury-based medicines. Going even further back in time, Greek, Roman, and Carthaginian smelters were big polluters. They put thousands of tons of lead into the air, causing fallout worldwide.

Technology's impact on the environment is dramatically seen in the extinction of animal species, the case of the passenger pigeon being a remarkable example. Once the most abundant land bird in the world, a single flock could hold a billion birds and shade the sun for hours as it passed overhead. One shot might bring down half a dozen birds, and even a thrown stick was an effective weapon. Once diners developed a taste for the bird, hunters used the telegraph to inform others about groupings and used the railroad to reach them. The population of passenger pigeons declined in proportion to the growth of railroad lines. The slaughter increased dramatically after 1850, and the pigeon was practically extinct within fifty years.

With many technologies, some environmental impact is unavoidable, and in each instance society must decide how much is acceptable. It is important to remember the long history of technologies' effect on the environment as we struggle with today's challenges, such as pollution and global warming.

Health

The nice thing about living in London
is you can see the air you're breathing.
—OSCAR WILDE

Finding the correct balance between technology's pros and cons has also been a challenge for centuries in the area of health. Consider the time before vaccines and antibiotics. The Black Death (1347–51) killed about one third of the population of Europe, perhaps twenty-five million people. Because of this pandemic, England lost as many as one thousand villages—so many people had died that the few survivors abandoned them. The impact would have been negligible without the technologies of cities that spread the disease locally and transportation that spread it remotely. More than 150 years passed before the population of Europe returned to its pre-plague level. Fewer workers meant higher wages and a weakened feudal system, laying the foundation for the Renaissance.

As significant as plague was in ending the medieval period, it also played a role in its beginning. Plague reached Constantinople in 542 CE and swept through Europe in the next few years with the same vicious speed as the Black Death eight hundred years later. This Plague of Justinian was named after the last Roman emperor. The Empire was in decline at the time, and Goths and Vandals had sacked the western capital in Rome. Justinian ruled the eastern Roman Empire and had reconquered much of Italy, but the plague devastated the army as well as the civilian population and aborted this last attempt at reunifying the Empire. Without the plague, Western Europe might have maintained its connection with Roman civilization and been spared its descent into the Dark Ages.

The contrast between the public health conditions of the Middle Ages and today is dramatic. The wars of the twentieth century may be the first in which disease didn't kill more than combat. The increase in life expectancy for civilians has also improved dramatically. For example, Shakespeare's Juliet was only thirteen years old. Considering suitors at her age wasn't so surprising when she would have been more than halfway through the average life span of the day.

Commodities as basic as pure water were scarce at that time. A total lack of knowledge about how disease was spread produced deplorable sanitation conditions, which often meant polluted streams and rivers. With water quality doubtful, people often drank weak beer as a clean alternative. The first public water purification system was built in London only in 1829.

Even then, the battle was far from over. Several decades before the pioneering work of Lister and Pasteur, Ignaz Semmelweis in Vienna proved that if doctors washed their hands in a chlorine solution before delivery, maternal mortality fell from roughly 20 to 1 percent. Nevertheless, much of the medical establishment rejected his simple technique for years. A few years later in 1854, Dr. John Snow neatly stopped the spread of cholera in one area of London by removing the handle of the water pump whose source had been contaminated. His results were also ignored. Appreciation of the fundamentals of clean water and safe sanitation did gradually catch on, and by the late 1800s, English social commentator John Ruskin was able to note a new attitude: "A good sewer was a far nobler and a far holier thing . . . than the most admired Madonna ever painted." Improved health may be the most important gift technology has delivered.

Human Physiology

People have got to learn to live with the facts of life,
and part of the facts of life are fallout.
—WILLARD LIBBY, member of the Atomic
  Energy Commission (1955)

While many new technologies have imposed emotional strains such as anxiety and fear on the public, some of the most dramatic strains are *physical*. Evolution perfected the human body to live in a particular niche, but technology has imposed on the body new conditions it wasn't designed for, at a rate much faster than it can adapt to. Despite concerns that riding trains at the furious speeds possible in the early 1800s would cause bodily injury, speed alone doesn't cause problems. However, as astronauts and fighter pilots know, acceleration can. And as anyone who has survived a car accident knows, sudden deceleration can, too.

The ancient technology of sea travel caused what is probably the earliest clash of human physiology and technology, motion sickness. And motion sickness continues to be a problem in our most modern form of transport, space travel. It has also been an unexpected side effect of long-term virtual reality immersion. In small doses, however, the sensation of motion can be used for entertainment, as roller coaster riders can appreciate.

Humans were inadvertent guinea pigs in the mid-1800s when working in caissons to dig bridge foundations far beneath rivers. The abrupt drop in air pressure when workers returned to the surface sometimes caused decompression sickness (the bends) as excess nitrogen in the blood bubbled out of solution. The bends can be painful, debilitating, and even fatal. Washington Roebling, crippled by this disease after working in a caisson under New York's East River, could only watch from his window the opening festivities of the colossal structure he designed and supervised, the Brooklyn Bridge. The bends are also a problem for scuba divers, and only after years of experimentation do recreational scuba divers now have reliable procedures for safe diving.

The opposite problem is altitude sickness. Moving from sea level to elevations much above ten thousand feet requires acclimatization. This wasn't a problem when travelers moved by foot or by horse, but modern transport can make the transition happen much more quickly. If a person moves too high too fast, symptoms can range from mild to life threatening. Even with proper acclimatization, mountain climbers can't survive indefinitely at extremely high altitude.

Vertigo is another problem that technology presents to modern people. Leaning over a high balcony is about as close as most of us would care to come to the girder walking that high-rise ironworkers perform as part of their jobs, but Mohawk Indians have sought this work for over a century. They have participated in every major New York City construction project, including the Empire State Building, the World Trade Center, and the Verrazano-Narrows Bridge. As susceptible to vertigo as any of us, the Mohawks overcome their natural fears by determination and experience. Many people struggle with a similar anxiety when obliged to fly.

Jet lag is a nuisance of modern travel, but hints of it were noticed centuries ago. When the tattered remnants of Magellan's expedition returned to Spain in 1522 after a three-year voyage circumnavigating the earth, they were startled to discover that they had lost an entire day. They were the first travelers to experience this. The fictional Phileas Fogg made the opposite discovery as he traveled *Around the World in Eighty Days* in the other direction. It took aircraft with the speed and endurance to carry passengers through several time zones in a single flight to introduce jet lag, a more substantial clash between human physiology and technology, Lyndon Johnson's approach to jet lag was to remain on Washington time when he traveled, forcing local dignitaries to meet at his convenience (don't try this unless you're the president).

Safety

Hell in a harness
—DAVY CROCKETT's evaluation of the railroad

Train accidents have as long a history as trains themselves. One observer at the opening of the Manchester & Liverpool Railroad in 1830, at the beginning of the Railway Age, commented: "The folly of seven hundred people going fifteen miles an hour . . . exceeds belief." The first train catastrophe happened in 1842 when over fifty people died in a crash near Paris. Train accidents in the United States increased in severity soon thereafter. By the 1850s, accidents began causing double-digit fatalities—forty-six due to an open bridge in 1853, thirty-four from a head-on collision in 1854, and sixty after a bridge collapse in 1857. Train fatalities could be expected every two or three months. Newspapers wrote sensational and often exaggerated accounts of the wrecks. Disaster stories have been selling papers for centuries.

Steamboats, much bigger than trains, also had bigger accidents. In April 1865, the riverboat Sultana was steaming up the Mississippi River. Grossly overloaded with thousands of newly released Union soldiers from the notorious Andersonville military prison and straining against the current, the boat's boiler exploded. More than fifteen hundred men died. Boiler explosions were rare on trains, but perhaps because boats' boilers were large and required custom engineering, they were a major cause of disasters.

Steam's dangerous reputation bedeviled the early years of Thomas Edison's electric generator projects, even though his steam engines were at a safe remove from the public. Edison had built electric power equipment that delivered direct current, while the Westinghouse Company had developed superior alternating current technology. To cast doubt on the Westinghouse approach, Edison's engineers developed the first electric chair—using the competition's alternating current, of course. Before the verb *to electrocute* became accepted, Edison proposed "to westinghouse." With news of capital punishment by electrocution in the press, he asked (referring to alternating current), "Is this what your wife should be cooking with?" Gas companies saw electric bulbs displacing their gas lights and jumped into the debate. They warned of accidental electrocutions while Edison responded with bulletins containing grisly descriptions of gas explosions.

Fire has plagued buildings ever since wood was used to build them, but it became a greater threat when kerosene stoves and lamps were introduced into homes. Kerosene was a versatile new fuel that replaced more expensive whale oil in the 1860s. However, oil refining of the time was often careless, and the 1870s saw five thousand deaths per year due to poor-quality kerosene. The producers themselves weren't immune from safety hazards, either. Less than a year after the first oil well was drilled in Pennsylvania, its derrick burned down, as periodically happened to its neighbors. Safety signs posted in oil fields during the 1860s read: "Smokers Will Be Shot."

Particularly dramatic catastrophes have shaped public opinion. When fire broke out in the Triangle Shirtwaist Building one day in 1911, workers on the ninth floor discovered that the exits had been locked. The 146 dead were mostly young women, and the disaster raised public awareness of sweatshop working conditions. Fire in Chicago's huge Iroquois Theater in 1903 killed 600—more than that city's Great Fire of 1871. The theater had been billed as fireproof. However, the disaster did stimulate the passage of new safety laws for public places. Nuclear waste and nuclear power-plant safety is a burden modern society is still struggling to deal with. One Chernobyl citizen noted after the 1986 disaster, "At first I was frightened, but now that I have heard all the explanations, I am still frightened." Nuclear weapons are even more frightening, symbolized by the "Doomsday Clock" on the cover of the *Bulletin of the Atomic Scientists* magazine The clock has shown the symbolic number of minutes before midnight (nuclear war), varying from two minutes before midnight in 1953 when the United States and the Soviet Union both tested thermonuclear weapons, to seventeen minutes before in 1991, when both countries pledged significant cuts in their nuclear weapons arsenals.

During the 1920s some consumers actually sought radioactivity. Radium water was sold as a health tonic, a curious but dangerous fad. In those pre-Hiroshima days, radioactivity was seen not as dangerous and debilitating, but as healthful and energizing. Sales dropped after the widely publicized case of a business executive who had consumed over a thousand bottles of this highly radioactive product. He required surgery to remove much of his jaw and died a painful death in 1932.

The dangers of automobiles are well known, but let's look at their antecedent for a more complete perspective. Movies sometimes use the clip-clop of horses pulling cabs or delivering groceries to set the time frame to the Victorian era. The mood is one of pleasant, nostalgic busyness, but the reality was quite different. Edgar Allan Poe said that the sound from iron wheels and horseshoes on cobblestone streets was the best "contrivance for driving men mad through sheer noise." In New York City around 1900, fifteen thousand horses died each year from exhaustion, beatings, or accidents, and one million pounds of manure were produced daily. The dust from the manure was so noxious that windows were kept closed throughout the summer. While we understand the dangers of cars, we should also remember that the technology it replaced wasn't perfect either.

The popularity of the Ford Model T drove American family ownership of cars from 1 percent in 1910 to 26 percent a decade later, bringing an increased number of accidents along with them. When do you suppose the car speed record exceeded one hundred miles per hour?

1920? 1930? No—it was in 1904. Even at that time, many complained about the obsession with speed.

While cars take their toll a few people at a time, civil engineering failures can cause huge and well-publicized disasters. One afternoon in May 1889, an earthen dam on the Little Conemaugh River failed. Unfortunately, the volume held behind the dam was *not* so little, and more than four billion gallons of water poured down the narrow valley in a tsunami at times seventy-five feet tall. Johnstown, Pennsylvania, lay in its path. The flood swept away most of the buildings in town and piled debris, house fragments, and victims against a railway bridge. As a final irony, stoves in the wreckage set the debris aflame that evening, and many who had survived the flood died in the fire. The death toll was more than two thousand.

Nor are civil engineering disasters new. The Babylonian king Hammurabi, who ruled in the eighteenth century BCE, is famous for a code of justice that specified damages for these disasters. Similar to Semitic "eye for an eye" justice, it specified that if a house collapsed and killed the owner's son, the builder's son should be put to death as a penalty. If a dam failed and damaged crops, the dam owner should be sold into slavery to repay the loss. While this approach may seem quaint, remember that construction and irrigation were high tech four thousand years ago.

Safety problems (and more important for our analysis, *perceptions* of safety problems) due to technology have a long history. Disaster was acknowledged and weighed in Hammurabi's scales. Dams have collapsed and bridges have fallen; ships have sunk and boilers have exploded. Trains, initially viewed with alarm, were accepted slowly. Fire, one of our most ancient technologies, has never been truly harnessed. Given these precedents, perhaps coexistence with technology is inherently uneasy.

A fortune-teller once told a young man that he would be both poor and miserable until age forty. Eagerly, he asked what changed at that point. She replied, "You'll just be poor. By then, you'll be used to it." Society's response to many technology problems evolves this way—we just get used to them.

Risk

One doesn't discover new lands
without consenting to lose sight of the shore for a very long time.
—ANDRÉ GIDE, author (1926)

Sometimes big risks pay off: the Eiffel Tower and the Brooklyn Bridge, the Panama Canal and the transatlantic telegraph cable, the Apollo program and the ARPAnet (later, the Internet). And sometimes they don't. The Swedish flagship *Vasa*, built in 1628, was a formidable warship. It had been recklessly built, however, and it sank in the harbor before its maiden voyage. Henry Ford and Daniel Ludwig failed with their ambitious Amazon plantations. The Iridium satellite phone project used satellites in low orbits to provide worldwide telephone communication; unfortunately, the customers who could pay for it had better alternatives. Just a few years after it became operational in 1998, the project was sold for less than 1 percent of the $3.4 billion that had been invested in it.

Today, society is risk averse, but that luxury has been paid for with centuries of risk taking. Today's risks usually involve money, while those in the past often involved lives. Aviation pioneers pushed the boundaries, making airplanes today a safe form of transportation. Mining and farming today are dangerous occupations, but nothing like they were in the past when productivity was more important than safety. Factory workers and machine operators have far safer conditions today than their nineteenth-century counterparts, who had few laws to protect them. A sailor's life was particularly dangerous, and stories from fishing ships, cargo ships, and whalers tell of hardship as well as adventure. One extreme example is that out of five ships and about 250 men in Magellan's voyage, only one ship with 18 men returned. Magellan himself was one of the casualties.

Ongoing risks with a track record (such as the number of deaths per year) are easy to compare with each other. Nevertheless, most people weigh risks poorly. The average American is much likelier to die in a car accident than a plane crash, much likelier to die from lightning than fireworks, and much likelier to die from influenza than anthrax. You're less likely to win the jackpot in a major lottery than to die in an accident while driving to buy the ticket. The likeliest calamity that could happen to a traveler to another country is not terrorism or kidnapping, but a car accident. Tornadoes and hurricanes combined aren't as deadly as heat waves; heroin and cocaine combined aren't as deadly as alcohol. Risk experts say that nuclear power is quite safe and swimming is not, while most people feel the opposite. Money spent on disease research is only vaguely proportional to each disease's impact. We worry about cell phones and brain cancer when we should be worried about cell phones and driving. The way the public ranks fear doesn't match up with the real risks, and a technology that has the same death rate as a natural event is perceived as more dangerous.

Accurately judging the likelihood of different outcomes doesn't come naturally. For example, imagine a football or soccer game. Now imagine collecting all twenty-two players on the field and asking them to compare birthdays. What are the chances that any two will share a birthday? Surprisingly, finding a match is almost an even bet. Throw in a referee and it's more than likely. Since most people don't work with probabilities enough to become comfortable with them, our society at large tends to grossly distort the threat of different types of risks.

No application of technology is universally acknowledged as solely good. Take genetically modified foods as an example. Some say that the good outweighs the bad, but others disagree. Do improved yields, better quality, or reduced pesticide demand outweigh the risk that new genes might escape into the environment and cause problems? Or, take hormone-treated cattle: The beef is cheaper, but are there health risks? Or, power plants: Is a coal-fired plant worth the downside of acid rain, pollution, open-pit mining, and other environmental problems?

For centuries, society has given technology the benefit of the doubt, assuming it to be innocent until proven guilty. A very different approach puts the burden of proof on a new technology and judges it guilty until proven innocent. That is, a technology must be proven safe before it is adopted. This philosophy is in vogue in Europe, to the annoyance of U.S. exporters of genetically modified foods. This Precautionary Principle can be seen as an extension of the initial words of the physician's Hippocratic oath: first, do no harm.

Medical Ethics

The real problem is not whether machines think
but whether men do.
—B. F. SKINNER, psychologist (1969)

Dr. Nancy Wexler helped develop a genetic test for Huntington's disease, a degenerative brain disorder that gradually but inevitably progresses to death. Her interest was more than scientific: she had a 50 percent chance of carrying the rare Huntington's gene herself. Nevertheless, she declined to take the test because there is no cure for the disease. She labeled this dilemma the Tiresias complex after the Greek prophet who observed, "Wisdom without benefit brings only sorrow." Is there value in knowing the future—in this case, good health or a terminal disease—when nothing can be done with that information?

In perhaps no area has technology affected our daily lives more than in the field of medicine. Although the changes have been, for the most part, dramatically for the better, some results raise ethical and legal questions. For example, is it acceptable to clone animals, and if so, what about people? When can people choose to end their own lives? Should a baby born with deformities be helped to survive even though that life would be short and of poor quality? How much should be spent to prolong the life of an adult with a terminal disease? Can a baby conceived after the father has died, using his frozen sperm, claim inheritance rights? Can a person sell an organ such as a kidney to someone who needs a transplant? If patients with tuberculosis refuse to take their medicine, can society cure them involuntarily for the public good? Does a surrogate mother have any rights to the baby that she carried? Is abortion ethical?

These ethical challenges are just the most recent ones, and we often forget those of the past. As medicine began to change how and when people died, it encroached on religious and spiritual beliefs, then as now a sacred part of many people's lives. Is it right to use antibiotics or vaccinations to change the natural course of a person's life? Are medical students justified in dissecting a corpse? Is contraception allowable? Is medicine more effective than prayer, and is illness actually *not* divine punishment?

When asked what he thought about Western civilization, Mahatma Gandhi said, "I think it would be a good idea." On Gandhi's list of seven major blunders of the world was science without humanity. Ensuring science *with* humanity will be an ongoing challenge, but don't think that it is a new one. This debate has already been going on for centuries.

Don't be afraid to take a big step when one is indicated.
You cannot cross a chasm in two small jumps.
—DAVID LLOYD GEORGE, British prime minister

11 Fear and Anxiety

IMAGINE AN INFANT FROM MEDIEVAL TIMES transported to the present day. Would that child grow up holding medieval customs and beliefs? Of course not—he would grow up just like his twenty-first century peers. Physically, we are no different than the illiterate and superstitious peasant living a thousand years ago. Knowledge has advanced quickly, much faster than evolution has adapted us to this new world. We face present-day realities with psychological and emotional needs unchanged from the Middle Ages—or even the Bronze Age.

This may explain the attraction of fringe science like astrology. What seems commonplace today was often amazing or even shocking when first available. Understanding each new technology takes time, and we fill in the gap with erroneous stories of how technology works. Some technology fears are valid (electricity *can* be dangerous), but even in this age of science, people manufacture their own.

Incredulity and Naiveté

The actual realization of the astonishing fact,
that instantaneous personal conversation can be held
between persons hundreds of miles apart,
can only be fully attained by witnessing the wonderful fact itself.
—a Rochester, New York, newspaper commenting on
   the anticipated arrival of the telegraph in 1846

Residents of towns newly connected to the telegraph network often showed amazement, disbelief, and even fear of the new technology.

Some refused to walk under the telegraph wires. In a challenge similar to the man versus machine contest between John Henry and the steam drill, one farmer bet that his team of horses could outrun a telegraph message. In 1842, just two years before the first working telegraph system, Senator Oliver Smith wrote after witnessing a demonstration, "I watched [Samuel Morse's] face closely to see if he was not deranged, and was assured by other Senators as we left the room that they had no confidence in it either." While some things must be seen to be believed, apparently others must be believed to be seen.

Most of our reactions to technology are the modern equivalents of the pride, enthusiasm, and fears felt by people of earlier centuries. We're not the first to be astounded by technology, not the first to rearrange our lifestyle to accommodate technology, not the first to wrestle with ethical dilemmas created by technology, and not the first to ask ourselves if technology's good outweighs its bad. Even our Information Age had precedents. Amazement at today's technology isn't unique— it isn't even particularly substantial.

When Alexander Graham Bell demonstrated the telephone in 1876, his invention received good reviews, but one newspaper wondered if "the powers of darkness are somehow in league with it." It is hard for us to imagine living in a society being changed by technology so dramatically that supernatural forces are suspected to be the cause.

If dark forces are imagined behind the telephone, one wonders at the reaction twenty years later to the ghostly images made with X-rays. At about the same time, the first movies were shown publicly. One presented a scene at the seashore—no monsters, no invading army, just waves rolling in along a beach. The crowd was terrified. They ran from the makeshift movie theater to escape the onrushing water.

The initial success of the 1858 transatlantic telegraph cable was marked throughout the country by parades and other celebrations. In a generation, news that had traveled only as fast as horses or boats could carry it began appearing in daily newspapers a day after it happened—just as it does today.

Charles Babbage, inventor of the first automatic digital computer in the mid-1830s, recalls his difficulties in conveying the new idea of the computer. "On two occasions I have been asked [by members of Parliament], 'Pray, Mr. Babbage, if you put into the machine wrong figures, will the right answers come out?' I am not able rightly to apprehend the kind of confusion of ideas that could provoke such a question." Just twenty years ago we could probably have found people similarly confused.

Technology advanced so fast in the 1800s that many observers couldn't keep up. In the autobiography *The Education of Henry Adams,* the author describes his reaction to the 1893 Columbian Exposition in Chicago. He found himself psychologically overwhelmed by the huge machines on display: "Probably this was the first time since historians existed that any of them had sat down helpless before a mechanical sequence." The reactions to some of these new technologies show how jarring they were. Is there a modern equivalent? Perhaps only news as unsettling as proof of aliens or ESP could produce a similar reaction.

In 1906 Lee De Forest designed the triode, a vacuum tube that made radio possible. While raising money for his projects, he inadvertently made enemies. Legal documents from 1913 charged that "De Forest has said . . . that it would be possible to transmit the human voice across the Atlantic before many years. Based on these absurd and deliberately misleading statements, the misguided public . . . has been persuaded to purchase stock in this company." Absurd? Human speech was indeed transmitted across the Atlantic just two years later.

Today, as in the past, people can have a hard time keeping up with technological developments, and advertising sometimes preys on people's confusion. One of my favorites is a modern equivalent of a pitch for patent medicine (concoctions backed by nothing more than the pitch man's smooth talk). By 1989 cable and satellite TV competed with local over-the-air broadcasts. An ad at that time in a national newspaper insert offered an ordinary TV antenna dressed up to look like a tiny satellite dish. The ad accurately stated that it didn't use cable or satellite broadcasts, then stated what it *did* do in such breathless terms that it sounded like a new innovation: "Works entirely via proven 'RF' technology—actually pull signals *right out of the air.* Instantly locks onto every local VHF and UHF channel from 2 to 83 to bring you their movies, sports, and special events *just like an ordinary pair of 'rabbit ears'*" (italics in original). It performed this miracle because it *was* an ordinary rabbit ears antenna. The ad concluded with this summary: "Not technical razzle-dazzle but the sheer aesthetic superiority of its elegant parabolic design makes the GFX-1000 a *marketing breakthrough!*" In other words: there are no interesting features here, but because of its unusual appearance, we'll sell a boatload of these things. The information is accurate, but the ad writers must have been laughing out loud imagining whom they would trap with this one. This is the equivalent of ads for stimulants or weight loss products that state "guaranteed placebo."

Arthur C. Clarke observed, "Any sufficiently advanced technology is indistinguishable from magic." Amazement at technology is hardly unique to our time, and history shows us many fundamental and sometimes even magical technologies.

Fears and Weird Beliefs

We don't see things as they are,
we see them as we are.
—ANAÏS NIN, author

President Benjamin Harrison had electricity installed in the White House, but family members were so unnerved by the first electric chair execution in 1890 that they rarely turned off the lights. Early users had difficulty understanding what electricity could do safely. Unwarranted fear reappeared in another form in the 1960s. Though accustomed to the safe use of electricity, some users saw electricity from nuclear power plants as being somehow more dangerous than that from fossil fuel plants. Change seen accurately is scary enough, but people often burden themselves with additional technophobia.

Every era has its own peculiar fears that reflect the issues of the day. At one time casino gambling was the new addiction and mental attack from aliens was the new delusion. Now, wasting time online is the new addiction, and mental attacks are perceived to come through the Internet. Fears and delusions can change with the technology. Apprehension about witches and unnamed dangers in dark forests, in vogue in the 1700s, has been replaced by concerns over crop circles and alien abductions. Updated delusions aren't just a Western phenomenon. For example, witches in Bali had traditionally assumed the shape of flames, animals, or treacherous women when seeking victims, but they are now imagined in modern shapes including driverless motorbikes with tires that pulsate as if breathing.

Fears of famine extend back to the dawn of agriculture and before. In the 1960s and '70s, we heard predictions of famine because of the population explosion. Food production has actually increased 40 percent faster than the population. The fears that population would outstrip food production were well founded, though wrong, and famines today are not due to a worldwide lack of food. Nevertheless, gloomy predictions still find a ready audience. Denis Dutton observed: "The steady evaporation of the question, 'When will overpopulation create worldwide starvation?' has left a gaping hole in the mental universe of the doomsayers. They have been quick to fill it with [other anxieties]. There appears to be a hard-wired human propensity to invent threats where they cannot clearly be discovered."

Despite public education and the increasing importance of technology in our daily lives, fringe science thrives. Not all people demand scientific proof for surprising claims, and there are plenty of believers in numerology, superstitions, healing magnets, UFOs, parapsychology, Atlantis, and the TV psychics who claim the ability to talk with the dead. Many of us have a need to believe.

Some people think that the Apollo moon landing was a hoax, and some give credibility to a chain letter's threats of bad luck to the person who breaks the chain. Some on the edge of poverty think that a lottery ticket is a wise investment, and others pay $2.95 a minute for advice by phone from a fortune-teller. Others think you catch a cold by getting cold, and almost every daily newspaper includes a horoscope. Though few still believe that spirits cause illness, responding to a sneeze with "Bless you" remains a holdover of etiquette. Many hotels don't have a thirteenth floor, and there is a Las Vegas hotel without floors forty to forty-nine to satisfy Japanese gamblers who consider the number four unlucky. The end of patent medicines in the 1930s hasn't meant an end to empty cures offered to gullible consumers. Robert Park in *Voodoo Science* observed, "There are, unfortunately, few scientific claims so far-fetched that no Ph.D. scientist can be found to vouch for them."

Our medieval minds can also fall prey to hoaxes. Eight-year-old Virginia O'Hanlon noted that "If you see it in the *Sun*, it's so" when she asked for advice, and the *New York Sun* replied with its famous, "Yes, Virginia, there is a Santa Claus" editorial in 1897. But this rock of credibility came from more flexible beginnings. Two years after its launch in 1833, a weeklong series of amazing stories gave the *Sun* the world's largest circulation. Their little secret: the stories were completely fabricated. The paper had focused on sensational news from its beginning, but the Great Moon Hoax told of lunar observations made by a powerful new telescope that discovered strange new plants, animals, and bat-like people. A few years later, the *Sun* ran an Edgar Allan Poe hoax about a successful transatlantic balloon crossing. So much genuine progress was happening at this time that stories like these seemed plausible.

The Cardiff Giant, apparently the fossilized remains of a ten-foot-tall man, was unearthed on a New York farm in 1869. Soon after the finders had been well compensated for their discovery, it was shown to be nothing but a clumsy hoax. Even then, visitors paid to see it—and still do today. Orson Welles' *The War of the Worlds* radio play, complete with disclaimers, panicked thousands during its broadcast on Halloween, 1938.

There may be undocumented animals in remote parts of the world, but pranksters don't help when they plant Bigfoot footprints and invent Loch Ness monster photos. And aliens may have visited Earth, but crop circles and many UFO "sightings" are hoaxes. It's hard for the rest of us, as outsiders to the worlds of cryptozoologists and UFO hunters, to identify any truths amid data that is known to contain many frauds.

I remember a newspaper photo claiming to show the offspring of a dog and a cat (it was titled "Is dat so?"). Years before the first cloned animal, I was taken in by an article claiming the birth of an elephant/mammoth hybrid using genetic material from a frozen mammoth. Since then, I've learned to check the date on a sensational news story to make sure it's not April first. You may have noticed that e-mail with a bogus virus warning is itself a virus, propagating itself through people instead of computers.

"Sales" of the Brooklyn Bridge and the Eiffel Tower to unwary rustics have a modern counterpart that brings us to the present. Through 2003, when its founder was jailed, the company Lunar Embassy sold plots of land on the moon to investors in the United States and Europe. These weren't sold as inexpensive novelties but for the hefty price of $1,600 each.

Keeping in mind society's illogical side helps when seeking to understand its reaction to technology. For centuries we've had difficulty separating technology truth from fiction.

The Adult Learning Curve

Conversation can be easily carried on after slight practice
and with the occasional repetition of a word or sentence.
After a few trials, the ear becomes accustomed to the peculiar sound.
—early Bell Telephone advertisement (1877)

In 1971 the tiny Tasaday tribe was discovered in a remote area of the Philippines. Its members lived in the jungle under Stone Age conditions. This anthropological discovery, considered one of the most important of the twentieth century, prompted a TV special, books, and international aid.

A great story, but this one, too, appears to be a hoax. Nevertheless, the idea of primitive people introduced to our world is fascinating. How would they react? Immersed in our culture as we are, all but the newest technology is largely invisible to us. Outsiders can give us a new perspective. Let's look at a few examples of people being abruptly introduced to new technology to remember how it feels to be a novice.

During World War I, the British Army recruited thousands of Gurkha soldiers. Ships were unfamiliar to these men from remote, landlocked Nepal. How did this huge thing move? Where were the legs or the rails? Almost four centuries earlier, the huge sailing vessels of Ferdinand Magellan's expedition awed Pacific islanders. Some natives thought the rowboats were the children of the galleons. American Indians faced the railroad with its noisy, smoking trains, and they saw people magically communicate over great distances with the telegraph. A balloonist, landing in a Spanish village in the late 1700s, was hailed as a saint. Another balloonist was assumed to be a witch. The Alaska oil pipeline brought money to Inuit communities, but this has often meant too-easy access to alcohol, televisions playing shows about a foreign culture and in a foreign language, and boredom due to the loss of traditional chores.

During World War II the U.S. military brought huge quantities of clothes, food, fuel, and weapons to Pacific islands that had never seen such goods. With the end of the war, the planes stopped landing. To induce the planes to return and bring more supplies ("cargo"), a cargo cult formed based on their simple ideas of cause and effect. Islanders noticed how planes were directed to land by ground personnel. Thinking that this was the only cause, they made their own mock headphones and antennas and lit signal fires to attract the planes.

Conversing by telephone is completely natural for us, yet the process was so miraculous and startling that some early users were incapacitated by stage fright. Early ads had instructions on how to use a telephone. Non-English speakers asked if the telephone could transmit their language. Some older adults grew up when a long-distance telephone call was reserved for short, important messages, and they still see it that way. They are unable to relax and chat when they envision a meter running.

These examples remind us to be tolerant of those struggling with today's technology. Perhaps you have heard stories of new computer users who, when told during the installation process to "type any key," complained they couldn't find the "Any" key. Or users calling PC support to complain that their screens are blank during a power failure. One user asked for a replacement for the PC's broken "cup holder"—or at least for what he had been using as a cup holder. It turned out to be the CD drive tray. When floppy disks were still floppy, one user identified them by attaching blank labels and then rolling the disks into a typewriter to label the contents. Another reported that the floppy software always worked the first time, but only the first time. After much investigation, the support person discovered that after using the floppy, the user attached it to the side of a filing cabinet— with a magnet. From writing, to ships, to telephones, to computers, people have needed time to adjust to new technology.

Skill Loss and Encapsulation

The discovery of the alphabet will create forgetfulness in the learner's soul,
because they will not use their memories;
they will trust to the external written characters
and not remember of themselves.
—SOCRATES, *Phaedrus* (470 BCE)

Socrates continues: "You give your disciples not truth, but only the semblance of truth; they will be heroes of many things, and will have learned nothing; they will appear to be omniscient, and generally know nothing." The Greek alphabet had been developed a few centuries before Socrates, yet his comments show that it was not yet universally accepted.

The memory skills he feared for were impressive. For example, Homer's *Iliad*, an important part of Greek oral history, was sung or recited from memory over perhaps five long evenings. Or, consider the Greek poet Simonides, who stepped out of a banquet just before the roof collapsed. Legend says that he was able to give rescuers the names of each of the several hundred people who had been inside, including where each had been sitting.

Writing records only the words, not the delivery. It does not capture components of the spoken word such as emphasis, volume, pacing, and pauses that are so important to the actor or oral historian. Gestures and expressions are also missing. Adults can read a children's story with a monotone or with different voices for each character, and either approach is valid from the standpoint of the text. To take a musical analogy, having only one recording of Beethoven's Ninth Symphony would not satisfy music lovers because every conductor and orchestra has a different interpretation—each gets different results from the same written score.

This rejection by Socrates of technology's obvious benefits seems ridiculous today, and yet the widespread availability of calculators in the 1970s raised similar concerns. Should the technology be embraced when it might erode important math skills? The consensus was that students could use them, but only after learning the fundamentals of math so they would know what a right answer looked like.

Perhaps even this condition will be relaxed. Traditionalists may eventually appear as old-fashioned as Socrates when calculators can know what problem is being solved and do the work themselves. Take another example: suppose word processor spelling and grammar checkers, which are nice tools but no substitute for a strong knowledge of English, someday become as good as an English teacher. Will teaching spelling become less important in response? We have ceded drudgery to machines in the past, and they will off-load new burdens in the future.

This concern over skill loss is connected to the concept of *encapsulation*. To illustrate this concept, let's take the example of a car with an automatic transmission. The gears have been hidden—they are encapsulated—and you don't worry about shifting them. Encapsulation is also what makes urban children think that food comes from the grocery store.

Society is filled with examples of encapsulation. Unlike the early days of radio, when hobbyists tinkered with their home sets and learned how to fix them, home electronics appliances often now have a warning sticker forbidding the user to open it. Computers have made the same migration: from home-built kits in the mid-1970s to desktop computers with card slots for expansion to laptops that mustn't be opened. Much that was serviceable in the past is now off limits, and components or devices are often discarded rather than fixed. Electricity is not yet too cheap to meter, but many appliances *are* too cheap to repair.

In the early days, drivers were not only welcome to tweak and adjust their cars, they were obliged to. Cars have become much easier to operate, with the encapsulation of the gearshift into the automatic transmission, and of the choke and spark advance into the engine's automatic controls. To take a boating example, maneuvering a power-boat is easy, but maneuvering a sailboat takes practice. Few understand the "Outer Limits" TV show reference, "*We* control the horizontal; *we* control the vertical," when TVs no longer need manual controls for tint, color, and horizontal and vertical sync. Medieval scribes needed to be able to make parchment and ink, and Renaissance artists needed to mix pigments; contrast this with buying paper, ink, and paints at the store today. Anybody who owns a camera can record landscapes or portraits without being an artist. Music is much easier to make when "play" refers to a button on a CD player rather than a musical instrument.

Around 1910 a Bell Telephone statistician projected that every working-age American woman would be needed as a switchboard operator within twenty years at the current rate of growth in that profession. This vast increase in operators didn't happen, of course—or maybe it did. According to the definitions of 1910, the new dial telephone and automatic switching technology allowed *every caller* to connect a call and do the job of the switchboard operator. Perhaps we could also say that every computer user does the work of a computer programmer—using the definition of the 1970s. At that time, it took a programmer to make a hobbyist PC do anything useful, but the wide range of packaged software now available gives that ability to the average user.

Ever more technology is made available to the user, and yet the burden often doesn't increase. We learn new skills but ignore old ones—how to repair a car or radio, draw or paint, play a piano or sing. Encapsulation doesn't just *cause* skill loss, it *enables* skill loss. The philosopher Alfred North Whitehead in *Introduction to Mathematics* (1911) observed, "[It is wrong to say] that we should cultivate the habit of thinking what we are doing. The precise opposite is the case. Civilization advances by extending the number of important operations which we can perform without thinking about them."

Encapsulation isn't a new phenomenon—it came with civilization. The specialization that the growth of cities enabled is a form of encapsulation. The farmer can focus on growing food, leaving the blacksmiths, soldiers, tanners, priests, artists, scholars, and other specialists to provide society's other needs. Abilities are encapsulated over time and become inaccessible, either physically (you can't get to them, like the components inside a chip) or intellectually (you take the ability for granted yet can't perform it yourself, like farming or textile making).

We humans have a limited ability to handle complexity, and encapsulation is the mechanism that allows life to avoid becoming ever more complex. Complexity is put inside a black box, giving the user a simpler interface. When that interface is later made complex with new features, capabilities can again be collected and put behind the scenes.

Today, examples of skill loss are often disparaged in ways such as, "Kids today can't spell" (or perform math or use grammar or whatever). But each example of skill loss will eventually not matter. Like Socrates, we must gracefully give up skills to technology as necessary.

A great many things are possible but not practical.
—ISAAC ASIMOV, author

12 Technologies That Touch Us

THE RECENT HISTORY OF DESIGN offers an interesting account of society's relationship to technology, alternately embracing it and becoming nostalgic for simpler days. At the beginning of the 1900s, the Arts and Crafts school was a reaction against the move toward machine-made products. Designers were concerned that quality was being discarded in favor of quantity and low cost. Art Deco became influential in the 1920s and streamlining was a frequent design element. Not only cars but consumer products, such as refrigerators, bicycles, toasters, and even pencil sharpeners, began to look like airplanes. This school embraced new materials, such as stainless steel, aluminum, and plastic, and moved from a plain to a more ornamental appearance.

After the austerity of World War II came biomorphic designs that rejected the machine lines of the 1930s. Flowers inspired skirts, giving them waists representing vine stems. Round, blobby shapes were common in household products. Concurrent with this were rocket-inspired car designs with big tail fins and chrome, and architecture's "International Style," from which came the principle "form follows function." This trend gave us the boxy buildings of gleaming glass and metal, a simple but bold celebration of the materials from which they are made.

Lots of technology—shipping, factories, and power plants, for example—affect us only indirectly, but many technologies touch us personally. These include home innovations, language, and timekeeping. While easily ignored, they are some of our oldest technologies and are still changing. We now turn to technology's personal side.

Home Innovations

I have a microwave fireplace.
You can lay down in front of the fire all night in eight minutes.
—STEVEN WRIGHT, humorist

Though insatiable today, much of the demand for commodities such as oil and electricity had to be created. As John D. Rockefeller's Standard Oil strove for control of the petroleum supply, it also worked to increase demand. Kerosene, the primary output of refineries of the late 1800s, was a new product. What was it good for? Standard Oil answered that with a variety of essential new products sold near cost, including heaters, stoves, and lanterns, all of which used kerosene. As demand increased, they were ready to satisfy it.

Thomas Edison faced a similar problem. His product was electricity, and the early years of this industry gave us the refrigerator, electric iron, washing machine, electric clock, and vacuum cleaner. Similarly, altruism didn't drive Michelin to publish its restaurant evaluations; it started distributing its tourist guides in 1926 to encourage motorists to put more miles on their Michelin tires. Here, too, demand had to be created.

These new products helped create what has been called the industrial revolution in the home in the early years of the 1900s. In addition to the conveniences already mentioned, the stove now consumed electricity or gas instead of wood, faucets delivered water to sinks throughout the house, and the refrigerator reduced the frequency of shopping trips and kept food fresher. The impact of some home technology goes back even further. In 1859 one observer noted: "It is astonishing how, in a few years, the sewing machine has made such strides in popular favor, [going from] a mechanical wonder [to] a household necessity."

Capitalizing on the reliable mail and railroad infrastructure, companies such as Montgomery Ward and Sears, Roebuck were able to offer rural families, without convenient access to stores, mail order catalogs that became second in importance only to the Bible in many farmhouses. Montgomery Ward, the earlier player, began in 1872 and by 1893 had a catalog of 544 pages. Even groceries were available. Today's e-commerce is only the most recent step in a long line of home-shopping innovations.

The design and construction of the house itself is also quite old. The wood-frame design used for most American houses today was a great improvement over the timber-frame construction technique. It used standard instead of custom-made components, reduced the need for skilled labor, cut the building time required, and used much less wood. It was developed in the early 1800s and hasn't been greatly improved on since.

Of course, not all innovations aimed at the home are successful. Monsanto's "House of the Future" at Disneyland opened in 1957. The entire shell—floors, walls, and ceilings—was made of plastic, their view of life in the 1980s. Thomas Edison proposed a poured concrete house in 1902. Designed as low-income housing, it could be built in four days for $300. Buckminster Fuller's 1929 Dymaxion House was prefabricated with aluminum sheets and sold for $1500. Fuller also developed the geodesic dome in 1947. Though very practical in certain applications, domes as homes have never been more than curiosities. Polyurethane foam had its proponents in the 1970s. Plastic, concrete, and aluminum sheeting have yet to form a major part of the structure of the modern house. Nor has steel, titanium, or carbon fiber; in the twenty-first century, we're still building houses mostly out of wood.

Consider the difficult path the metric system has followed. Metric is compelling, it is already taught in school, and it costs little to implement. New metric package sizes could be easily phased in since food packaging is updated frequently. The United States was an early advocate of decimalization and the first country to adopt a decimal-based currency (in 1792), and Congress legalized the metric system in 1866. Scientists and engineers use it as a rule, but Americans still have not discarded their comfortable but clumsy English units in daily life. Among the nations of the world, the United States shares this status with only Liberia and Burma.

Electronics has had more than its share of flops in the home. Robert Lucky's use of a Picturephone at Bell Labs in the early 1970s sounds like the poignant end to a science fiction story: "I think I had the last one in the world. Alas, there was no one left to call." The videophone is an excellent reminder that technical challenges aren't the only ones that must be overcome.

Despite the overheated enthusiasm around new products, not all succeed. Some, like plastic homes and the Picturephone, find interest but no buyers. Even the successful ones often face resistance.

Timekeeping

And we conjecture [that Gulliver's pocket watch]
is either some unknown Animal, or the God that he worships . . .
[for] he seldom did anything without consulting it.
—JONATHAN SWIFT, *Gulliver's Travels* (1726)

The Pulitzer Prize-winning book *Soul of a New Machine* (1981), by Tracy Kidder, relates the story of the building of a new Data General computer. The work was challenging and the hours punishing. One hardware engineer, tired of the frustrations of digital components that switch in billionths of a second, quit and left this note: "I'm going to a commune in Vermont and will deal with no unit of time shorter than a season."

Timekeeping has a long history, as has resistance against it. The second-century BCE Roman dramatist Plautus griped, "Confound him . . . who in this place set up a sundial to cut and hack my days so wretchedly into small portions!" Even though the use of sundials was an important advance in timekeeping, they aren't especially accurate and only work when the sun shines. Water clocks, hourglasses, and mechanical clocks provided ever-improving accuracy. In fact, from the 1300s through the early 1900s, clock accuracy doubled every thirty years, imposing exponential improvement in timekeeping technology just like Moore's Law would later impose on semiconductors. A highlight of this progression is John Harrison's series of ingenious chronometers. To calculate a ship's longitude, they provided the essential element of accurate time. By 1761 one of his clocks was accurate to within one second per week, even when carried on a rolling ship.

Pre-Industrial Revolution workers didn't have convenient access to accurate time, but that didn't matter. They could deduce the time accurately enough for most purposes by looking for simple cues: sun position, feelings of hunger or sleepiness, and so on. With the rise of factories, things changed. Time regulated the factory's operation, though watches were expensive and scarce and only the foreman would have one. Control of time meant power.

Appreciation for the importance of time was increasing, but gradually—as the case of *Hadley v. Baxendale* illustrates. In this 1854 English legal case, a mill was forced to shut down because of a problem with a part. The faulty part was shipped away for repairs, and because the factory couldn't operate without it, managers specified that the part be sent by train. However, it was shipped instead by canal, a cheaper but slower route. The factory understood that time was money, but this was not obvious to the shipper. They sued for wasted time but lost the case because the judge ruled that *damages must be foreseeable.* The concept of time urgency was taking hold but wasn't yet universally appreciated.

Affordable pocket watches in the mid-1800s made accurate time accessible to the average citizen. But time can be a harsh master. Some doctors worried about the medical problems caused by fallible people trying to obey infallible timepieces. Office workers were not only able to get much more done, they were required to. Stress-related disease increased dramatically in a generation, and one researcher of the period stated that never before did inventions "penetrate so deeply, so tyrannically, into the life of every individual."

Henry David Thoreau in *Walden* (1854) noted how trains had become the heartbeat of rural America: "[Trains] go and come with such regularity and precision, and their whistle can be heard so far, that the farmers set their clocks by them, and thus one well-conducted institution regulates a whole country." He asked, "Have not men improved somewhat in punctuality since the railroad was invented? . . . To do things 'railroad fashion' is now the byword." Thoreau had identified the 1850s' version of "Internet time."

Before trains there was no standard definition of local time, and municipalities could adjust it anyway they wanted. Towns could define their noon to be the same as that in a neighboring town, noon could be defined as mean solar noon, or it could be something else entirely. For a town to define time in a way that was convenient for them worked well when they were remote from one another, but not when railroads connected them. For the first time, a person could travel fast enough to be hampered by the illogical time differences between places. A coast-to-coast trip in 1870 crossed perhaps a hundred time zones. Not only was it inconvenient for the railroad to juggle the different times to ensure that trains arrived when passengers expected them, but collisions were becoming more common as train engineers became unsure about which time applied where. To improve safety and convenience, the railroad industry created a simple nationwide standard with four U.S. time zones in 1883. This was the predecessor of the worldwide time standard adopted the following year.

While railroads helped define time, electronic communications helped broadcast it. Britain had a nationwide time signal sent by telegraph in 1852, and the rest of the industrialized world soon copied this service. At a time when clocks were still bought from jewelers, these retailers often subscribed to the telegraph time service. Passersby were encouraged to drop in to set their watches, which spread the correct time throughout society.

Time information began to be wireless by about 1900. The Eiffel Tower was nearly demolished in 1909, just twenty years after its construction and at that time still the tallest structure in the world, but its utility as an antenna to send radio time signals saved it. These signals helped ensure to-the-second accuracy over increasingly wide areas. After phone companies provided a time service, a significant fraction of all calls were for the time. Perfectly accurate time has been available to U.S. consumers for almost a century. In 1916, they could buy electric clocks that took time information from the power lines. This put on power companies the burden of ensuring that, on average, their power had exactly sixty cycles per second.

GPS (Global Positioning Satellite) is the latest version of accurate timekeeping. Originally designed for the military and started in 1993, GPS is widely used for consumer applications. Extremely precise atomic clocks on satellites broadcast timing information, and the differences in how long the signals from different satellites take are used to compute the location of the receiver. GPS has made accurate time a worldwide commodity. In an interesting historical echo, GPS provides accurate time as a secondary benefit of its location-computing role, just like Harrison's chronometer of close to 250 years ago. We take accurate, convenient time for granted, often ignorant of the tremendous advances made in previous centuries.

There is more to time than just accuracy, and a French innovation from two hundred years ago illustrates this. Comedian Steve Martin told of a visit to France and confided in the audience: "They have a different word . . . for *everything!*" After the French Revolution, they even had a different way to tell time. Republican time defined a day of ten hours, each containing one hundred minutes of one hundred seconds, creating a new second with about the same length as our own. While the metric system defined during the same period is still with us, Republican time was just a short-lived experiment.

Only familiarity with our conventional timekeeping system prevents us from marveling at how ridiculous it is. Seconds and minutes are in base 60 (inherited from the Babylonians), hours are in base 12 (or maybe 24), but everything is represented in base 10. No wonder children have difficulty learning it. Any other system would be more logical.

A modern experiment in decimal time is called (what else?) "Internet Time." Proposed by Swiss watchmaker Swatch, Internet Time divides each day into one thousand beats. There are no hours or minutes, there are no time zones, there is no daylight savings time, and there is no A.M. or P.M. For example, the time @854.17 (854.17 beats after midnight in time zone GMT+1) is the same worldwide regardless of the local time. Proponents argue that not only is using base 10 for time superior to the crazy set of bases used by conventional time, but people from different time zones can agree on meeting times or airplane departure times with an unambiguous time reference. Will Internet Time have more impact than French Republican Time? Only time will tell.

Calendars

Hurry has a clearly debilitating effect upon the tissues
and may in time injure the heart.
—BRITISH DOCTOR (early twentieth century)

The aftermath of the French Revolution brought a new calendar as well as a new method of timekeeping. This calendar still had twelve months, but arbitrary features were made logical. The year started on the autumn equinox, roughly September 22. New ordinal names were given to the days: *Primidi* (first day), *Duodi* (second day), *Tridi*, and so on. The month names within each season rhymed; for example, the summer months were *Messidor* (harvest month), *Thermidor* (hot month), and *Fructidor* (fruit month). Every month had thirty days and, in a deliberate move to de-Christianize the calendar, weeks were gone and each month was composed of three *décades* of ten days each. The year's five extra days (or six, in the case of a leap year) had separate names, were not part of any month, and were placed at the end of the year. The experiments with Republican Time and the Republican Calendar ended during Napoleon's reign.

This calendar briefly returned a feature lost with the introduction of the Gregorian calendar: that each date falls on the same day every year. For example, July 28 can fall on any day of the week, but 10 *Thermidor* always fell on a *Decadi*. As with many new technologies, our Gregorian calendar wasn't superior to its predecessor on all counts. In fact, when it was introduced to correct errors that had accumulated in the sixteen hundred years since the Julian calendar began, many saw it as an imposition rather than an innovation. Ten days had to be dropped to put the calendar back in phase with the solar year. Monthly salaries or rents in countless situations had to be reconsidered when the month lost so many days. One would think that the calendar war was over by 1582 when the Gregorian calendar was first adopted—or at least by 1752 when the English-speaking world also made the switch. But only after China fully adopted the Gregorian calendar in 1949 did the majority of the world follow it. Even now there are dozens of calendars still in use and active proposals to improve the Gregorian calendar. The Y2K problem and the question of when the twenty-first century really started are only the most recent clashes with a technology that has regulated our days since the earliest civilizations.

Writing

You had to make them awful interesting at that price or get fired.
—ERNEST HEMINGWAY, commenting on newspaper articles
that cost $1.25 per word when sent by telegraph

King Sejong of Korea ruled in the early 1400s when Koreans wrote with Chinese characters. He wanted an independent written language for Korean and so created Hangul, a character-based script, to replace pictograph-based Chinese. According to legend, once the system was developed, King Sejong faced an additional challenge. Buddhist monks held great power at that time and imposing Hangul without their support would have been impossible. To address this the king first painted the characters with honey on leaves and let ants eat them away. He then summoned his religious advisors to interpret the leaves. They postulated that the marks represented a divinely inspired new alphabet. Duly empowered, the king made the alphabet official.

Technology can constrain language. A commentator in the early days of the telegraph observed, "The delicacy, intricacy, and nuance of language is endangered by the wires." Some say that Hemingway's terse writing style came from his early years as an international correspondent, reporting over the telegraph where every word counted. Newspaper stories had to be short and unambiguous, and they drifted toward a plain and uniform style, regardless of the author.

Bandwidth over wires is no longer a problem. However, e-mail has replaced some categories of business letter, and e-mails are often terse, like telegrams before them. Another constraint on writing is handwrting recognition software, and voice recognition software can force slow and deliberate speech. Writers in many European languages sometimes find their accented characters converted into gibberish by programs or character sets that only understand English letters. Where this is a problem, many writers are constrained to use equivalents—*aa* for *å* in Danish, *ss* for *ß* in German, and so on. Marshall McLuhan observed, "We shape our tools, and then our tools shape us."

Constraints imposed by technology are usually temporary, but they're also perennial. Those ushered in by the telegraph are gone, but now we have new ones. They are like weeds: as we work to eliminate those we have, others will sprout.

Let's remember also the cultural impact made by explorers, settlers, and missionaries from the late 1400s onwards, particularly in Africa and the Americas. As much as half of the world's six thousand languages are no longer spoken by any children. Cultures are being homogenized. Transportation and communication technologies can connect and teach, but they can also fragment and destroy.

Technology Words

Modern English is the Wal-Mart of languages:
convenient, huge, hard to avoid, superficially friendly,
and devouring all rivals in its eagerness to expand.
—MARK ABLEY, journalist

Once the Internet became an important consumer utility many new names were proposed for it. Some were grand, like "Information Superhighway" or "National Information Infrastructure"; others were shorter and more practical, like "I-way" and "Infobahn." Finally, consensus returned to where it had started, and we now use some variation of *Internet* or *Web*. We may be close to the next step in their evolution: the use of lower-case letters, as *internet* and *web*.

Society also groped for names for older technologies. For example, "flying machine" wasn't the only precursor to *airplane*. Others included "aero-motive engine" and "aerial velocipede." Early names are often built from words people already understand. We know what the telegraph does, so if something conveys the same Morse code through the air, it's "wireless telegraphy." A machine that replaces the horse for travel is a "horseless carriage." We still don't have a single word to refer to a general road user—a car or truck driver, or a bike or motorcycle rider. To remedy this, "roadent" was proposed. Other misses have been "picture radio" (television), "iron horse" (railroad), and "optical engine" (telescope). This archaic use of *engine* to mean a device that accomplishes something, as in "analytic engine" for computer, is present in the modern term *search engine*.

There are plenty of new words and phrases for modern technology. Some endure (*road rage*) and some don't: a "leadite" for a person who prefers writing with pencils over PDAs, "cuddletech" for cute technology, and "jetiquette" for airplane etiquette. Words are somewhat like technology itself as far as characteristics that lead to broad adoption: a new word is tried and, if it serves a purpose, it is used by more and more people; we gradually become accustomed to a new word and then use it without thinking about it; and younger people are more eager to adopt new words, leaving older generations as late adopters.

Words to denote new technologies usually aren't completely new. *Railroad, telegraph,* and *e-mail* may refer to completely new things, but the words are built from known components. Sometimes, old words find new service. *Broadcast* initially meant to cast seeds widely, which is a nice metaphor for spreading an electronic message. New definitions have also been pressed onto *surf, browser,* and *cloverleaf.*

Words also go in the other direction. Technology-only phrases may eventually be used to refer to something outside their original domain. My son once referred to the time between waking and getting out of bed as *booting up.* We might investigate the *flip side* of an issue, or we might make new contacts by *networking.* The boss might put my project *on hold,* or perhaps *pull the plug* on it. An impatient person might say that the *meter is running,* and a hidden issue might be *under the radar.* Someone might need to *let off steam.* From electrical engineering, we have *turn on* and *live wire;* from radio, *tune in* and *on the same wavelength;* and from nuclear engineering, *critical mass* and *ground zero.* These are all used in contexts different from the original technological one, where insightful people found new analogies for existing technology phrases.

What do you say when you answer the phone? Alexander Graham Bell preferred "Ahoy" or "Hoy" and used this greeting for the rest of his life. Others were tried, such as "What is wanted?" and "Are you ready to talk?" and even "Are you there?" "Hello" was considered undignified, and the proper form was debated for decades. When people met in the late 1800s, protocol often required different greetings based on relative social standing. What can you say when it's your turn to talk but you don't know to whom you are talking? The telephone made social class invisible.

Some words came from technology, but have drifted so much that their original meanings can be surprising. *Trivia* came from the Latin for "three roads"—a meeting place where written news was put. *Vellum,* which now usually refers to a high-quality paper, has the same origin as *veal:* before paper, the treated skin of juvenile animals, such as lambs, calves, and kids, were used for writing. *Manufacture* originally meant "make by hand." Even the word *computer* has an archaic origin. For centuries, it referred only to human calculators. During World War II, scientists working on the Manhattan Project created an assembly line of human computers. Each worked on a single step of a large problem with slide rules, tables of numbers, or electromechanical calculators, effectively creating a bio-electro-mechanical computer. The word evolved to mean an electronic device used for fast computing, but that meaning, too, is out of date. Today word processing and games are more important than computing mathematical constants or ballistics data.

Technology impacts vocabulary quite visibly in the area of "retronyms," words that technology has redefined. Fifty years ago, all clocks had hands. Now, to refer to a clock with hands, you specify an *analog clock.* The word *television* is now synonymous with *color television,* producing the retronym *black and white television.* To avoid confusion with the electronic edition of a newspaper, you might refer to its *print edition*; to avoid confusion with a paperback, you might specify the *hardcover book.* Technology has created other retronyms: acoustic guitar, regular coffee, real cream, natural food, cloth diaper, paper notebook, manual transmission, conventional oven, biological virus, film photography, live operators, and surface mail. There's even *book-book* (that is, not an e-book) and *wood-wood* (a golf driver made of wood).

The etymology of many words documents the impact of the technology of travel. When explorers came across a new thing, they usually tried to duplicate the natives' name for it. We got our words *banana* and *yam* from Africa, *tomato* and *chocolate* from the Americas, and many words from the Caribbean, including *canoe, maize, hurricane, potato, hammock,* and *tobacco. Raccoon* came from Algonquin, *gecko* from Malay, and *kangaroo* from an Australian Aboriginal language. From Turkish came *sherbet* and *coffee*; from Arabic, *sash* and *alcohol*; from Chinese, *ketchup* and *kowtow*; and from Hindi, *jungle* and *pajamas.*

With the dramatic influence French had on English after the Norman Conquest in 1066, you might think that French would comfortably accept useful English words in return. But the august *Académie Française* (French Academy) has been minimizing this type of erosion in the purity of French since 1635. The Académie and the French government have thoughtfully provided substitutes for foreign technology words that have crept into the language. These substitutes are encouraged for all French speakers but are mandatory in official contexts. However, excellent substitutes aren't possible in every case. For example, *courrier electronique* is the clumsy replacement for *e-mail*.

The vocabulary of modern technology is being cautiously incorporated into other languages, also. The German post office long insisted on labeling its telephone booths *Fernsprecher* (remote-talker), a word with German roots, and resisting Greek-based *Telefon*, the word in popular use. They yielded to the public consensus only in 1981. In Iceland, before glass was available, the amniotic sac from a cow (*skjár*) was stretched to make a translucent covering for a window. *Skjár* became the word for window, whether organic or glass. When a word was needed for a computer monitor, this same ancient word was pressed into service. The Icelandic word *tölva* means computer and derives from the words for *numbers* and *prophetess*. Would we see computers differently if we called them "number-prophets"?

Standardization of Language

Language was not made by man,
but rather the other way around.
—Francisco Varela,
  biologist and philosopher

The early 1700s was a time when science was defining temperature, distance, colors, and sounds. Shouldn't language be similarly reliable? Jonathan Swift, Daniel Defoe, and other writers of the time pushed for standardization of both English meaning and spelling. The earliest important English dictionaries were published during this period.

Even with dictionaries' stabilizing influence, however, words continue to change meanings. Samuel Johnson, whose famous dictionary was published in 1755, rejected the suggestion of any English equivalent of the French *Académie Française* or the Italian *Accademia della Crusca*. He said that lexicographers should register the language, not embalm it.

American Sign Language (the fourth most-used language in the United States) has more in common with sign language in France than in Britain. Here was a great opportunity missed—even in sign language we have no universal language.

Movies initially spoke a universal language by speaking almost no language at all. Silent movies were understandable by just about everyone. European immigrants who couldn't understand English could enjoy them, as could native English speakers watching non-English imports. Then talkies fragmented the market.

Looking back to Latin, however, we see a real lingua franca. Latin was the main language of learning throughout Europe up to the time of the printing press, although Europe was as polyglot then as now. The increasing body of literature printed in local languages undercut Latin's dominance. The Church fought the rising importance of vernacular languages: it saw its role as a gatekeeper and was not enthusiastic about parishioners being able to access the Bible themselves. William Tyndale's 1525 New Testament was the first printed English translation. Forced out of England by the Church, he printed it in Germany. He was captured and executed a decade later. The Church's harsh response is odd since the traditional Latin version of the Bible was itself a fifth-century translation from the original Hebrew, Aramaic, and Greek texts.

English, in particular, had been considered a vulgar tongue but was gradually seen as a language that could support the debate of great ideas. Sir Isaac Newton lived through this transition period, and while he wrote his *Principia* (1687) in Latin, his *Opticks* (1704) was in English.

In order to create (or *re*-create) a world where a single language is understood by most of civilized society, hundreds of artificial languages have been proposed. These invented languages are potentially as versatile as natural languages like French or German but much easier to learn. In the 1600s, some proposals were inspired by math to replace words with numbers or symbols. The mathematicians Descartes and Leibniz suggested early versions along these lines. Solresol was a language that used the seven musical notes (do, re, mi, fa, so, la, ti), instead of letters. It had a long run of support and is the only language that can be played on a musical instrument as well as spoken. By the 1800s, most proposals were streamlined amalgams of existing languages. Esperanto (launched in 1887) has had the most success, and some children have been raised speaking it as their first language. While English has 728 irregular verbs, Esperanto has none—and just sixteen grammar rules.

Predictions for Esperanto and other languages have been grand. Proponents believe that giving everyone a common second language would be an important step toward world peace. While the goal of dictionaries was to bring together those speaking a single language, that of artificial languages is to bring *everyone* together. At present, the front-runner for the universal second language is not an artificial language at all—it's English.

Units of Measurement

The most profound technologies are those that disappear.
They weave themselves into the fabric of everyday life
until they are indistinguishable from it.
—MARK WEISER, Xerox computer scientist (1991)

Measurement standards have adapted to keep pace with technological advances. The English system that the United States inherited is a simplified form of a somewhat arbitrary accumulation of units: we have 12 inches in a foot, 3 feet in a yard, and 1760 yards in a mile. There are ounces, teaspoons, furlongs, and acres. There are avoirdupois weights and troy weights and apothecary weights. And this is just the English system; other countries have their own homegrown sets of units.

The metric system finally created a logical system of units. It has grown so that there is a unit for every need, and decimal prefixes such as micro (one millionth) and mega (million) easily allow measurements from the tiny to the huge. You may never have heard of a *megameter,* but it's easy to figure out that it's a million meters.

And yet, while many English units seem arbitrary, there is regularity that may not be obvious. Units often use the binary (base 2) number system, the one used by computers. In volume measurements for example, there are two gills (now archaic) in a cup, two cups in a pint, two pints in a quart, two quarts in a pottle (also archaic), and two pottles in a gallon. Dry measures (such as peck, bucket, and bushel) also form a binary system. The binary system is a natural way to express divisions: a quarter stick of butter (rather than 0.25 sticks of butter, if we were to express this as a decimal), an eighth of an inch, or even half a meter.

We can see that the English measurement system isn't always arbitrary, and it turns out that the metric system, strained with new needs brought about by the widespread use of computers, is missing an important element in the area of binary measurements that the English system has worked with for centuries. The metric system is routinely misused for measurements of bytes. For example, a hundred-gigabyte hard disk does indeed hold a hundred billion bytes, as expected, but a PC with one gigabyte of memory (RAM) actually has about 7 percent *more* than one billion bytes. In this example, the *giga-* prefix has been hijacked to mean the binary value 2^{30} (1,073,741,824) rather than its intended 10^9 (1,000,000,000). New prefixes have been proposed by the International Electrotechnical Commission to provide the needed vocabulary. Using their prefixes, our memory example above would have been accurate if it had referred to one "gebibyte"

(1 GiB), not one gigabyte (1 GB) of memory. The new prefixes are Ki (kebi-) for 2^{10}, Mi (mebi-) for 2^{20}, Gi (gebi-) for 2^{30}, and so on.

The metric system's insistence on having every measurement in decimal turns out to be a liability here. The creaky old English system of units is more in tune with the Computer Age than we might have expected, and the two-hundred-year-old metric system may need an update.

> *Technology: . . . the knack of so arranging the world*
> *that we don't have to experience it.*
> —MAX FRISCH, writer (1957)

13 Innovation Stimulation

ALMOST TWO THOUSAND YEARS AGO, Hero of Alexandria made a toy called the *aeolipile*. It is a small sphere that spins on a hollow axle. Steam, forced through the axle, exits the sphere through two jets pointing in opposite directions. The force of the steam causes the sphere to spin. While designed to be a novelty, this is thought to be the world's first steam engine.

The Romans were masters of engineering, building roads, coliseums, aqueducts, and other impressive projects. With the aeolipile, they had the insight that ultimately led to the Industrial Revolution seventeen hundred years later. Our lives today might be vastly different if the Industrial Revolution had been ignited in the first century CE.

Why didn't the Romans pursue steam power? The likeliest explanation seems to be that they had no motivation. Slave labor was an important component of the Roman economy, and idle slaves were dangerous. Laborsaving inventions could have been social suicide.

Different forces drive the industries that give us our new technologies. Slavery discouraged innovation in the Roman case, but gold rushes (literal and figurative), prizes, and exploration of new lands have long been stimulators of business and innovation. We will consider these driving forces and will also examine the Industrial Revolution as an extreme case study of innovation stimulation.

Gold Rush

A blind alley paved with gold
—ECONOMIST ANATOLE KALETSKY,
  critiquing the 1990s technology boom

Imagine searching for Easter eggs in a dark room. Now, imagine searching for Easter eggs in a dark room in which there are *no* Easter eggs. Finally, imagine searching for Easter eggs in a dark room in which there are no Easter eggs, after someone shouts, "I found one!" Add in billions of dollars to fuel the chase and that's what the late 1990s Internet boom was like as venture capitalists and investors scrambled to find the next Microsoft or Intel.

The Internet boom is often dated from the 1995 Netscape initial public stock offering. At the end of the first day of trading, Netscape was worth over $2 billion even though it had not made a dollar in profit. Other Internet companies soon made splashy public offerings, including Yahoo!, Amazon, and eBay. Stock prices for Internet companies soared. AOL bought Netscape, then paid $111 billion to buy the venerable media company Time Warner, creating AOL Time Warner. Day traders caught the enthusiasm and pushed up stock prices. Venture capitalists seemed to have more money than businesses to throw it at. It was a giddy time, and conventional wisdom only seemed to get in the way. Profit would take care of itself—"synergy" was the key, or maybe it was "getting eyeballs." Non-Internet companies joined in the frenzy by simply appending *.com* to their name and were rewarded with big jumps in their stock prices. A single domain (www.business.com) was bought for $7.5 million. The 2000 Super Bowl showed ads from seventeen dot-coms, at $2 million a pop.

A few months later the NASDAQ stock index peaked. The Internet bubble was over. Just three dot-coms advertised on the next Super Bowl. Seven of the seventeen from the previous year were already out of business. A few years later, after the largest corporate loss in history ($99 billion for 2002), the "AOL" was dropped from Time Warner's name. Hundreds of high-flying Internet companies were no more.

Was it an exaggeration to say that there were *no* Easter eggs (Internet home runs) in that dark room? Probably. There are a few large Internet companies that have old-fashioned characteristics such as profit, a decent business plan, and a respectable price-to-earnings ratio, though the overall result of that five-year gold rush disappointed almost every investor. Frenzied though the Internet bubble was, however, it wasn't unprecedented.

Consider a real gold rush as an example. The Klondike gold rush of 1897 illustrates the difficulty of striking it rich. According to the Klondike Gold Rush National Historical Park:

- 100,000 people set out for the Klondike gold fields
- 40,000 reached Dawson City
- 20,000 stayed to search for gold
- 4,000 found it
- 300 found enough to be considered rich
- 50 managed to keep their wealth

In other words, of the people who went looking for gold, only one person out of two thousand got what they wanted—or what they *expected*, in many cases. One of the few who made it home with a decent amount of money was John Nordstrom. He used his Klondike earnings to start a shoe store in Seattle that grew into the Nordstrom department store.

Samuel Brannan responded differently from most of those infected with gold fever. He was a shopkeeper at Sutter's Mill, California, when gold was found there in 1848. While a less perceptive man might have cashed in everything and staked a claim before the crowd arrived, Brannan instead traveled to San Francisco to spread the word of the spectacular find—after first buying every shovel in the city. Supplying goods to the miners made him California's first millionaire.

Who makes the big money during a gold rush? In most cases, it has been the owner of the ship, hotel, saloon, or store, not the miner. These intermediaries mine the miners. We saw the same thing with the Internet boom. The majority of revenue in the Internet industry went to the suppliers of equipment (such as servers and fiber optic lines) or software. These suppliers made a profit with each sale, while their customers made the long-shot bet that they could strike it rich with new businesses that consumers would care about. And as Internet access moves from competitive advantage to business necessity, the required equipment becomes simply a cost of doing business—good for the intermediaries but a burden for their customers.

Boomtowns are committed to a single industry and are hit hard when that industry fails. The United States has had its share of lumber, coal, gold, silver, and oil boomtowns and ghost towns. For example, Silver City, New Mexico, and Tombstone, Arizona, were built on silver, and Virginia City, Montana, and Nome, Alaska, were built on gold. Once places of opportunity, they are now little more than footnotes and tourist attractions.

Oil was discovered at tiny Pithole, Pennsylvania, in January 1865, six years after the world's first oil well was drilled nearby. In nine months, Pithole had fifteen thousand people, fifty-seven hotels, and a daily newspaper. Water pipes were laid in the main street. But within two years of the discovery, the population had dropped to two thousand because of well fires and falling production. A few years later, the town was back to its original size of a few hundred individuals, a ghost town of empty buildings and abandoned derricks in a polluted wasteland.

Perhaps the granddaddy of boomtowns is Potosí, thirteen thousand feet high in the Andes Mountains of Bolivia. Silver was discovered there in 1544 and exploited by the Spanish, swelling the town's population to more than 120,000 before 1600. Potosí was not only the biggest city in the Americas but one of the biggest in the world—it was bigger than Rome and half as big as Paris, then the largest city in Europe. At that time, the tiny settlements at Jamestown and Plymouth were still years away. Potosí dwindled along with the silver reserves over the next several centuries. Today it is a quiet regional capital.

Technology today doesn't lead to boomtowns like those that exploited gold or other resources, but technology downturns have left their mark. Office buildings sit empty, with a bankrupt company's name sitting forlornly on top and a "For Lease" sign planted in front. Boston's Route 128 experienced this in the 1980s as minicomputers took a hit, and the Bay Area's Silicon Valley experienced it after the Internet bubble.

Stock speculation has a long history, and the Internet stock bubble of the late 1990s is not without precedent. In the 1950s, when nuclear power was in vogue, having *uranium* in a company name gave it the cachet that *.com* would later have. During the electronics boom in the early 1960s, there was a corresponding boom in company names ending in *-tronics*. In the early 1980s, biotech was the industry that could do no wrong. When Genentech went public, its share price tripled in the first hour of trading.

Further back in time, revolutionary technologies such as canals, railroads, electricity, and radio triggered their own stock frenzies. Then, as now, a company with poor prospects might still have been able to sell its stock if it worked with the technology du jour. For example, electrical stocks were hot in London in 1882, a few years after the appearance of Edison's incandescent light. In a single two-week period, sixteen new companies went public. Tenuous claims were sometimes made to the hot technology of the moment. For example, Seaboard Airlines went public during the aviation bubble after Lindbergh's 1927 flight, but it was actually just a railroad.

Prizes

*"A true Englishman doesn't joke when he is talking about so serious a thing
as a wager," replied Phileas Fogg, solemnly.
"I will bet twenty thousand pounds against anyone who wishes
that I will make the tour of the world in eighty days or less."*
—JULES VERNE, *Around the World in Eighty Days* (1873)

By the late 1970s, the £50,000 Kremer Prize had stood unclaimed for almost twenty years. Created by British industrialist Henry Kremer, it required an aircraft under human power to navigate a half-mile-long figure eight. It was finally claimed by the seventy-pound Gossamer Condor, which now hangs in the Smithsonian Museum.

With that prize won, Kremer raised the bar: £100,000 for the first human-powered flight across the English Channel. Two years later, the Gossamer Albatross, built by the same team that had built the Condor, took this second Kremer Prize in 1979.

I eagerly followed these stories when they were news, and I think most people would agree that these successes were fascinating. But were they useful? Decades later, human-powered flight still isn't available to everybody who can ride a bicycle. I'm sure that even Henry Kremer himself didn't know what would grow from the research his prizes helped stimulate. But that's the point: if the benefits are large and obvious, there will be plenty of money pursuing them, from government or business. Unconventional or even eccentric prizes like these help drive innovation in areas that aren't already on someone's radar.

We've all heard of prizes such as the Pulitzer and the Nobel. There are many other prizes that are awarded after the fact to the best from a pool of excellent candidates—for example, the Millennium Technology Prize, the Japan Prize, and the Lemelson-MIT Prize, all for technology achievements. However, the Kremer Prize is in a completely different category. It looked forward; it defined and highlighted a target.

A more recent example in this category of forward-looking prizes is the Ansari X Prize. Established in 1995, the X Prize was for the first privately funded craft to fly twice into space (one hundred kilometers high) within two weeks. The $10 million prize was won by SpaceShipOne in 2004 on the anniversary of the 1957 launch of the Sputnik satellite. As Sputnik launched the Space Age, perhaps the flight of SpaceShipOne will reinvigorate it. Another recent prize was the $2 million DARPA Grand Challenge, a cross-country drive by autonomous vehicle from Los Angeles to Las Vegas, awarded in 2005.

Prizes have also stimulated research in nanotechnology. Richard Feynman focused attention on this area in 1959 and personally offered two $1000 prizes, one for a tiny motor and the other for tiny printing. Both have been won. Inspired by this example, the Foresight Institute offers a new $250,000 prize for nanotechnology innovation, named the Feynman Prize.

To stimulate research in computer science, the Electronic Frontier Foundation has offered a series of prizes for the discovery of large prime numbers. Researchers are closing in on the $100,000 prize for the first prime with ten million digits. The Loebner Prize offers $100,000 for the first computer program that passes the Turing Test, the ultimate test of artificial intelligence.

An offshoot of this is the making of a public wager. The Long Bets Foundation encourages and publicizes various disagreements on technology issues. For example, Mitch Kapor says, "By 2029 no computer—or 'machine intelligence'—will have passed the Turing Test." Ray Kurzweil disagrees. The two have wagered $10,000 apiece on this issue, with the money going to a charity of the winner's choice.

The history of innovation-stimulating prizes goes back a long way. The first transatlantic flight was made in 1919 by a biplane with a six-person crew. In the same year, Raymond Orteig, a wealthy French hotel owner, put up a $25,000 prize for the first nonstop flight between New York and Paris. Eight years later, it was won by Charles Lindbergh. His flight wasn't the first transatlantic flight, it wasn't the first *nonstop* transatlantic flight, and it wasn't even the first nonstop transatlantic flight between the *mainlands* of North America and Europe. In fact, there had been seven successful transatlantic flights by plane and airship before Lindbergh, carrying a total of roughly eighty people. But Orteig had picked the right goal. Not only did it catalyze nine teams to strive for the prize—which in total invested about fifteen times the amount of the prize—but the flight excited the public like none had before. The response to Lindbergh's achievement was extraordinary, and four million people lined the streets of New York to welcome Lindbergh home. Roughly one quarter of the country's entire population eventually saw him during his eighty-two-stop U.S. tour.

The Orteig Prize wasn't the only one in the area of flight. In fact, during the early years of the aviation industry, millions of dollars were offered through perhaps one hundred incentive prizes. And that first transatlantic flight in 1919? It was driven by a prize, too.

Other prizes go back further in time. A company offered a $10,000 prize for a new material that could replace ivory in billiard balls. John Hyatt responded in 1869 with celluloid, one of the first synthetic plastics. The invention of margarine won a French prize for a butter substitute in 1870. Napoleon Bonaparte offered a prize for a method to preserve food, won in 1809 with canning. The British Parliament created a £20,000 prize for a practical way to fix one's location at sea in 1714, and a century earlier the king of Spain had offered a similar prize.

Market forces aren't always enough to drive innovation. Sometimes high-profile (and lucrative) prizes are needed to catalyze the process.

Transportation and Exploration

Thus it appears that the sweltering inhabitants
of Charleston and New Orleans,
of Madras and Bombay and Calcutta,
drink at my well.
—HENRY DAVID THOREAU, *Walden* (1854),
  commenting on the export of winter ice
  from New England ponds

Well over a century passed between Columbus's first voyage (1492) and the Jamestown colony (1607). You'd expect such slow progress that long ago. But hold on—the European discovery of Brazil was six years after Columbus in 1498. Cuba was conquered in 1511. Two years later, Balboa crossed Panama to see the Pacific and Ponce de Leon landed in Florida. In 1519, Cortés conquered the Aztec empire in Mexico, in population roughly as large as Spain itself. The Magellan expedition was the first around the world three years after that, and in another decade, Pizarro conquered the vast Inca empire in Peru. In an instant of historic time, the technology of transportation changed the Americas forever—not to mention the pioneering work of the Portuguese to the east that culminated in Vasco da Gama's discovery of the sea route around Africa to India in 1498. Almost the full extent of the world (in coarse outline, at least) had been discovered in a generation.

Travel has been a tremendous driver of business, discovering and connecting suppliers, producers, and customers, and driving the demand for innovation. Compared to other technologies, travel has shown particularly unsteady progress. The Phoenicians sailed from the eastern Mediterranean to England for tin over three thousand years ago. After the unification brought by the Roman Empire, Western Europe became fragmented and lost its knowledge of the world. Only in the post-Columbian period did explorers begin again to push the boundaries and dispel the legends of dragons and unicorns, tribes of headless men with eyes in their abdomens, and the Asian empire of Prester John. Even within Europe, the quality of the Roman road infrastructure was not exceeded until perhaps 1600.

Outside Europe, exploration has also had setbacks. For example, Chinese voyages of discovery in the early 1400s ventured as far west as Africa. Compared to Columbus's expeditions almost a century later, these were huge—armadas of hundreds of ships and tens of thousands of sailors and soldiers. But with a new emperor came a new, isolationist policy, and the explorations stopped. Japan adopted a similarly introspective stance. Though it had more guns than any European country in the 1500s, Japan turned its back on outside influences for several hundred years until the 1850s.

Though the world is well explored in our own day, travel is still imperfect. War, bandits, and tribal boundaries slowed down a recent expedition that tried to follow Marco Polo's route through central Asia. Its members speculated that Marco Polo himself, protected by documents from the Khan, might have traversed the unified Mongol empire more safely seven hundred years earlier. And many of the islands of the Pacific are so remote that news of the end of World War II took decades to reach them. In the early 1970s, almost thirty years after the war's end, three Japanese soldiers hiding on three different islands finally surrendered.

Transportation in the United States is fairly stable today, but it has undergone periods of revolutionary change. Steamboats, the icons of Mark Twain's Mississippi, were used for only about sixty years. By the 1870s the railroad had made Mississippi steamboat travel uneconomical.

Like the Mississippi, the Oregon Trail was another important but short-lived route. Just a few years after the opening of this rugged two-thousand-mile trail in 1842, the Donner party left Illinois bound for California. Poor directions and bad weather trapped them in the Sierra Nevada Mountains as winter set in, when they were just short of their destination. Of the more than eighty people in the original party, only half survived, and many of these did so by eating their dead comrades.

Barely twenty years later, the transcontinental railroad enabled quick and safe trips to California. What had been a six-month trip now took six days, and the Oregon Trail was abandoned after only twenty-five years.

The era of the clipper ships was shorter still. These sleek vessels sacrificed payload for speed and were used for passengers or valuable cargoes like tea or spices. Though expensive to build and run, they sometimes repaid their costs in a single trip. They decreased in popularity after 1855, put out of business by difficult market conditions and competition from steamships. They had been influential for only a decade.

The Pony Express carried mail the two thousand miles between Missouri and California, cutting the three-week stagecoach delivery time in half. The transcontinental telegraph line in 1861 made it obsolete overnight. It had been in operation for less than two years.

By comparison, our own transportation system looks quite sedentary. Our most advanced means of commercial transportation, the Concorde supersonic airplane, was introduced thirty years ago and has since been retired.

The Industrial Revolution

We came here not to view your works
In hopes to be more wise,
But only, lest we go to Hell,
It may be no surprise.
—ROBERT BURNS, referring to Carron Iron Works
  (1787, modernized version)

The Industrial Revolution was a unique surge of innovation that came from an unprecedented cascade of inventions. Improvement in one area highlighted limitations in another. Pressure was then applied to address the new bottleneck, and so on. This process may sound familiar—like the logical, sequential improvements in a modern factory, such as an assembly line or chemical plant. But remember that there was no model to follow, no wise and experienced central commission guiding the process. The very nature of the modern factory was being invented at this time, and these changes were evolving in multiple industries across the entire country of England.

The Industrial Revolution initially burst forth within the textile industry. Weaving and spinning technology had been fairly stable for thousands of years (the Bible refers to weaving looms, for example), but that began to change in 1733. In this year, the flying shuttle was invented, the first shot of the Industrial Revolution arms race. This improved shuttle made a weaver roughly four times more productive.

The spinners were now the bottleneck. Entrepreneurs responded with the spinning jenny in 1764, which was basically six to eight spinning wheels working in parallel. Though much more productive, thread quality from the jenny was inferior. This led to other inventions by 1780 that produced high-quality thread and were powered by water, not human effort. To spin a pound of cotton into thread had taken five hundred hours by hand. Machines reduced this to twenty hours by 1780 and to just three hours a few decades later.

The weavers fired back with the water-powered loom in 1785. Then the carding of cotton was automated. Water power became insufficient, so steam power replaced it. With the tremendous increase in cloth making, cotton suppliers became a bottleneck. Cleaning cotton had taken a day per pound when done by hand, but the cotton gin (1793) increased productivity fiftyfold. By 1830, England had perhaps ten million spindles for spinning thread and over one hundred thousand looms, most powered by steam. One worker had become as productive as two or three hundred with manual equipment.

Like the trickle over an earthen dam that becomes a torrent, the change spread and grew. These technologies that worked so well with cotton were applied to silk, flax, and wool. The Jacquard loom wove elaborate designs with punch cards. The manufacture of stockings and lace was automated. Improvements were made in bleaching, dyeing, and printing.

Innovation spread from the textile industry into other industries. The manufacture of glass and pottery were automated. More demand for steam power meant more demand for coal, so coal mining ramped up in response. Tin, copper, and lead mining also expanded. In 1800, Britain imported iron; fifty years later, it produced more iron than the rest of the world combined and pioneered the production of cheap steel soon afterward. Thousands of miles of canals, followed by tens of thousands of miles of railway as well as steamship routes, connected mines to factories to markets. England had gone in a generation from a country like every other to a country like no other.

The privilege of being the first industrial nation was bought at great cost. As Alfred North Whitehead said, "The major advances in civilization are processes that all but wreck the societies in which they occur." England doubled in population in the fifty years ending in 1830, and many cities grew much faster than that. The living conditions in these industrialized cities were abysmal. Friedrich Engels, a manager at his father's company in Manchester, was shocked at what he saw and wrote *The Conditions of the Working Class in England* in 1845 to document and call attention to them. His firsthand account is the most important source for the following summary of the life of factory workers during this time.

By 1845, most factory workers were women and children. They could be paid less than men and were more compliant. This left most men unemployed, and many of these idle men turned to begging, drink, or crime. A typical home might be a cattle shed, or perhaps a cellar, often damp and sometimes flooded. These were always overcrowded, and furniture and even beds were a luxury. Diseases such as cholera were frequent and sometimes epidemic. Dust was everywhere (or mud, depending on the season), while clothes and shoes were scarce. Small children were often left at home when the older children and adults went to work, and these untended children often got into accidents. Laudanum (opium in alcohol) was commonly given to fussy children to keep them quiet.

Irish job seekers flooded into England at the rate of fifty thousand per year. Engels wrote, "The Irish have . . . discovered the minimum of the necessities of life, and are now making the English workers acquainted with it."

For six days a week in these cities the air was gray with the coal smoke billowing from hundreds of factory chimneys. Latrines, open sewers, pigsties, garbage, and tanneries added their smells. Rivers were also polluted; Manchester's Irk River was aptly named. It flowed into the city clean and transparent but left burdened with all kinds of human and industrial waste. Engels comments:

> [Beneath a bridge] flows, or rather stagnates, the Irk, a narrow, coal-black, foul-smelling stream, full of debris and refuse, which it

deposits on the shallower right bank. In dry weather, a long string of the most disgusting, blackish-green slime pools are left standing on this bank, from the depths of which bubbles of miasmatic gas constantly arise and give forth a stench unendurable even on the bridge forty or fifty feet above the surface of the stream. But besides this, the stream itself is checked every few paces by high [fences], behind which slime and refuse accumulate and rot in thick masses.

This filthy river was the only source of washing water in the poor parts of town.

Even the market was tainted. Meat was often spoiled or taken from sick animals. Commodities were often adulterated: sugar was mixed with powdered rice, coffee with chicory, flour with chalk, pepper with dust, cocoa with dirt. Tea was often mixed with other leaves or even with used tea.

The best that can be said about factory work is that it didn't require much effort. But the conditions were monotonous, noisy, and dusty; the work was meaningless; and the days were long. A worker's job might be to tend a machine and tie broken threads as necessary, or, perhaps, sharpen the points of needles over and over, day after day, and year after year. While this didn't require much physical or intellectual labor, it did require constant attention. When mishaps were due to workers' negligence, the overseer was on hand to fine them—or flog them in the case of children.

Accidents were common—some crippling, some fatal. Maybe a finger joint would be lost, or an arm or a leg. Even a simple cut might lead to tetanus, and many workers developed curved spines or legs. An injured worker would be lucky if his employer paid the doctor bills. If he were unfit for work afterward, it was his problem. Seeing the maimed people in the streets, Engels said, "It is like living in the midst of an army just returned from a [military] campaign."

Children were a large fraction of the workforce, initially because their small fingers were better able to tie broken threads, but later just because they were cheap. Children who should have been going to school or playing were instead working under the same conditions as most adults. Legislation was passed in 1847 that limited work to ten hours per day for six days a week. The bill applied only to women and children, and many manufacturers ignored it; still, conditions at this time were better than they had been. Laws in 1833 allowed fourteen- to eighteen-year-olds to work twelve hours per day and children ages nine to thirteen to work nine hours per day. Before this, factories saw children as young as five years old and workdays of up to sixteen hours. Social reformers who objected to these conditions were told that long hours kept children from mischief and adults from drink.

Note that this wasn't simply an unfortunate situation in some small corner of British society. It affected a huge fraction of the population. By 1800, 20 percent of the labor force worked in manufacturing or related industries. Fifty years later, this fraction increased to perhaps half.

Tension resulting from such working conditions and other changes brought about by the Industrial Revolution led to violence as far back as the mid-1700s when "machine breakers" attacked the inventors, mills, or machines that workers blamed for degrading their lives. But by 1811 things were different, and the attacks were now coordinated. For fifteen months, armed gangs of Luddites (followers of a possibly mythical Ned Ludd), terrorized factory owners in central England and destroyed textile machines. Parliament responded by making the destruction of a factory machine a hanging offense. In Luddite strong-holds, everything made of lead, including roofs and drains, was scavenged. People were making bullets, and rebellion seemed to be in the air. Despite the demands of the Napoleonic Wars, over ten thousand troops, supported by cavalry and artillery and about twenty thousand men in local militias, were mobilized to maintain order.

Military suppression and improved economic conditions ended the Luddite movement, but conditions were still grim. Even decades later, Engels feared that the workers' anger "must break out into a revolution in comparison with which the French Revolution . . . will prove to have been child's play."

Today, a Luddite is often considered to be someone against all technology, but this doesn't correctly characterize the original Luddites. Many were comfortable with it, having used manual weaving or spinning machines before being forced into factory work. Rather, they protested "all Machinery hurtful to Commonality"—that is, technology that damaged people and communities. This was more than a protest against the loss of jobs; this was a fight for a way of life. For centuries people had followed nature's cycles in small villages. Machines (if any) could be operated and maintained by a family. Uprooted and moved to filthy cities, people struggled to support themselves and raise their children in what was probably the most intense technology-induced social upheaval ever. The impact of our own Information Revolution has been hardly as dramatic as that from the Industrial Revolution—and for that we can be grateful.

Your Majesty, I have at my disposal what the whole world demands:
something which will uplift civilization more than ever
by relieving man of all undignified drudgery.
    —Matthew Boulton, speaking about steam power (circa 1770)

14 What's Mine Is Mine

THE COMMISSIONER OF THE U.S. PATENT OFFICE has been widely quoted as saying in 1899, "Everything that can be invented has been invented." How could he have said this at a time when the telegraph and railroad were maturing, the telephone and electric industries were rapidly growing, and the skyscraper and automobile were just beginning? The answer is that he *didn't* make the statement. The closest any patent commissioner seems to have come to this is speculation in 1843 that we might at some point reach an end of innovation—but only because of the furious rate of innovation at that time.

Intellectual property protection, such as patents for inventions and copyrights for books, music, and other creative products, has been an important driver of innovation since patents were first issued over five hundred years ago. The intellectual property issue most in the news at this writing is the use of peer-to-peer (P2P) sharing over the Internet. The Napster file-sharing service was an early P2P innovator. These networks have been judged as having substantial legitimate uses and are therefore legal—though their users may well break the law. Music studios have complained that these networks are causing them to lose sales, although the sales statistics are ambiguous and the actual impact is unclear.

Imagine if the VCR—which, it must be admitted, could be used to make copies of copyrighted broadcasts—was squashed before it got going. Too heavy a legal hand risks stifling the experimentation that drives breakthroughs and that benefits copyright holders as well as users. The challenge is to find the right balance.

Trade Secrets and Industrial Espionage

There is no country in the world
where machinery is so lovely as in America. . . .
The rise and fall of the steel rods,
the symmetrical motion of the great wheels
is the most beautiful rhythmic thing I have ever seen.
—OSCAR WILDE

Peter Chamberlen developed obstetrical forceps (used for delivering babies) in about 1630. He aided several English queens through childbirth, and his forceps gave him a substantial reputation and advantage over his competitors. Succeeding Chamberlen family members didn't share the forceps with other midwives, keeping the design as a trade secret. It was eventually made public after a few generations, but untold numbers of lives were lost in childbirth because of a century of self-interest.

In part to reduce trade secrets such as this, patents had first been issued in Europe in the 1400s. A patent is an exchange: the inventor gets a state-supported monopoly for a limited time and in return publicly reveals everything about the invention. There's no point in industrial espionage to uncover the details of a patented invention because those details are, by definition, publicly available. However, unscrupulous competitors have always had the incentive to steal trade secrets.

To break the Chinese monopoly on silk production, two Byzantine agents went to China to steal some of the precious silkworms in about 550 CE. Disguised as monks, they brought silkworm eggs and mulberry leaves back to Constantinople hidden inside their bamboo walking sticks. Not surprisingly, the Byzantines then did their best to maintain their own monopoly safe from their European neighbors.

Manaus, in the heart of the Brazilian rainforest, was a booming city in the late 1800s because of the growing rubber market it supplied. It built an opera house and a cathedral and was one of the first cities in Brazil with electricity. Even before its windfall began, however, the smuggled seeds of its downfall were being cultivated. Disease prevented the establishment of rubber plantations in Brazil, but no such problem was found in Malaysia. Malaysia's trees were mature enough to produce rubber by 1910. Rubber from Asia soon took over the market, leaving Brazilian sources (and Manaus) to shrivel.

Dutch smugglers brought coffee to Java around 1700, breaking a three-hundred-year-old Arab monopoly. The Spanish monopoly on cochineal, a brilliant and valuable red dye made from a New World insect, lasted for more than two centuries before it was lost to smugglers in the late 1700s. Even Thomas Jefferson stooped to espionage when he smuggled Italian rice to planters in the American South. He also paid to have seeds of a particular variety of hemp—valuable for making rope and canvas—smuggled out of China, though this was a capital offense.

Trade secrets have been even more important in areas other than agriculture. Some things we now think of as common knowledge were once carefully guarded secrets. Beginning in the 1400s, the logs made by ship captains documenting routes to new places and noting reefs, rocks, and safe passages were protected as company secrets. Often they were even state secrets. European maritime powers including England, France, the Netherlands, Portugal, and Spain all vied with one another to push the boundaries of the known world and uncover lucrative new trade routes.

China developed many technologies before the West, including iron casting, gunpowder, and paper, as well as china (true porcelain). Europe had pottery, of course, but china was different—it was light and translucent, in contrast to coarser European pottery. Interest in china grew along with the Chinese tea trade, and china was eventually carried to Europe by the ton. The first to re-create china in Europe was

Johann Böttger, who created Dresden china in 1707. So valuable was the secret that when Sweden invaded, Böttger and his assistants were moved to safety to keep the secret exclusive to Saxony.

Trade secrets were central to any craft occupation. Shipwrights used them to build clipper ships, Alfred Krupp in Germany used them to build his revolutionary steel cannon, and masons used them to build cathedrals. Samuel Slater memorized the details of Richard Arkwright's textile factory in England and returned to Rhode Island to build America's first factory in 1793. James Cabot Lowell did his own espionage in England and with that information built a mill in Massachusetts in 1814. Twenty years later, the Lowell mills produced close to one hundred miles of cloth *per day* and became the world's first integrated textile factory, handling every step of production from raw cotton to cloth.

Even now, patents are not the best way to protect all inventions. Coca-Cola guards its secret formula and Microsoft guards its source code. Employment agreements minimize the ability of an employee to bring trade secrets to a competitor. Soviet spies stole atomic secrets, and the United States in turn flew spy planes and satellites over the Soviet Union. The space programs of both countries got a boost from German rocketry experts after World War II. Trade secrets and espionage are alive and well today.

Patent Battles

Success. Four flights Thursday morning.
All against twenty-one-mile wind....
Average speed through air thirty-one miles.
Longest fifty-nine seconds. Inform press.
Home Christmas.
—WILBUR AND ORVILLE WRIGHT, telegram
   to their father from Kitty Hawk, NC,
   December 17, 1903

Orville Wright made the first successful flight in an airplane in 1903. Foregoing any public adulation, the Wrights carefully guarded their invention even as they continued to improve it. Though they allowed press coverage of that first flight, they waited more than four years to publicly demonstrate their airplane. The famous photo of the first flight, one of the most reproduced photos of all time, was deliberately blurred before it was published so competitors couldn't steal any ideas.

The Wrights had turned their prototype into a reliable aircraft by 1905, but they let Alberto Santos-Dumont make the first public demonstration in France in 1906 and let Glenn Curtis win a *Scientific American* prize for the first airplane to fly a one-kilometer course in 1908. Though they were eager for buyers, the Wrights demanded that a purchase be made sight unseen; only with a sales contract would they demonstrate their airplane.

Why were they so secretive? Since their invention would be visible to competitors, the Wrights had to protect it with patents. On the other hand, they knew that patents provided only modest protection. For example, they must have known of the blatant patent infringement that burdened Alexander Graham Bell. Telephone pirates operated by the hundreds in the less lucrative parts of the country, hoping to remain unobtrusive. They were like gnats—annoying but small and with safety in numbers. Nevertheless, the Bell Telephone Company fought six hundred lawsuits and won them all. Then as now, being guilty of patent infringement meant a fine, not jail time. A company might decide that copying patented technology was simply a risk worth taking and therefore a smart business decision.

Bell himself may have been on both ethical sides of the patent issue. The initial patent that Bell filed in 1876, on which his indispensable claim to priority rested, was filed just two hours before a competing claim by Elisha Gray. While this may seem like a remarkable stroke of luck for Bell, evidence suggests that he used bribery to uncover details of Gray's work and beat him to the punch.

Patents have been associated with no-holds-barred practices for centuries. By 1849 Alexander Bain of Scotland had perfected an automatic telegraph that would eventually send as many as one thousand words per minute, much faster than the manual telegraph. But his patents stepped on the toes of other inventors such as Samuel Morse, and patent battles drained his resources. Late in life, he was granted a small pension by the British government in recognition of his achievements, but by this point he had slipped into such poverty and obscurity that it was some time before he could even be located. He died a few years later in a home for incurables.

Almost every corner of technology has stories of patent battles, and some were especially nasty. Everyone has heard of Thomas Edison, but it was Nikola Tesla who invented the fundamentals of our electrical system: alternating current, dynamos, transformers, and motors. Tesla had worked for Edison, but the two had a falling out, and George Westinghouse was the beneficiary of most of Tesla's inventiveness. Edison responded with what some would call hardball marketing tactics (and others might call *dirty tricks*). Tesla died penniless in 1943.

Thomas Edison spent two million dollars over twelve years defending his rights to just one invention, the incandescent light. Edison also pioneered movie technology in his New Jersey lab, and his company held most of the industry's patents. California was a safer place to make movies for those who felt constrained by Edison's patents. By the time federal patent enforcement caught up with the lawbreakers, the patents had expired.

The invention of the television offers another example. Philo T. Farnsworth outlined the idea of television at age fifteen, applied for patents at age twenty, and went public with his invention in 1928 at age twenty-two. David Sarnoff of RCA tried to buy the patents, but Farnsworth wouldn't sell. Sarnoff retaliated by having one of his engineers reverse-engineer the Farnsworth design, then tried to bury Farnsworth in legal battles. Though Farnsworth won in court, his patents expired by the time TV sales took off. The two combatants both died in 1971, Sarnoff as a respected and wealthy visionary and Farnsworth as a forgotten and impoverished wretch.

Like Farnsworth, inventor Edwin Armstrong wanted to license his invention of FM radio rather than sell it. In 1954, frustrated by years of legal battles with infringing but deep-pocketed companies, he jumped from a window to his death.

A major reason why patent battles are common is that most of the knowledge necessary for a new invention is often available, with many inventors poised to make the breakthrough. True, we often credit a single person as the source of our fundamental inventions—Fulton for the steamboat, Morse for the telegraph, McCormick for the reaper, Bell for the telephone, Edison for the electric light, Marconi for the radio, the Wrights for the airplane, Berners-Lee for the World Wide Web, and so on. But if they hadn't developed their respective inventions, someone else soon would have. In many cases, someone else *did* invent it earlier but failed to create a successful product—which was the case with more than half of those in the list above.

Today we find heated patent debates, too. For example, are some software patents too sweeping? Should companies be allowed to patent genes? Is it ethical for a company to exist simply to buy intellectual property from others and then sue infringers, both real and imagined? Patents are as important today as ever, and as always, they are still used to define turf.

While the history of patents is full of stories of inventors fighting to defend against infringers, it also contains more cheerful stories of inventors who simply didn't bother to file for a patent or those who deliberately gave up their rights for the benefit of humanity.

- An independent inventor developed the first video game in 1958, fourteen years before the first commercially successful game. He said, "I considered the whole idea so obvious that it never occurred to me to think about a patent."
- AT&T judged the first use of its transistor, in a hearing aid, so commendable that it didn't ask for royalties.
- Jonas Salk said, "Who owns my polio vaccine? The people! Could you patent the sun?"

- John Muir was a talented inventor before he became a naturalist. His attitude was that "no inventor has the right to profit by an invention . . . really inspired by the Almighty."

Some of the developments given away freely have succeeded more than their developers could have imagined. Out of hundreds of competitors, the BASIC computer language became one of the most widespread. Another widely used computer development is TCP/IP, the networking protocol used for the Internet. After perfecting it, researchers at the University of California at Berkeley released it as open source (nonproprietary) software in 1992. Sir Tim Berners-Lee developed the first version of the World Wide Web. He won the one million euro Millennium Technology Prize in 2004, in large part because he didn't patent or commercialize his invention.

Armchair entrepreneurs have second-guessed this generosity. What if the developers of BASIC had gotten just ten cents for each BASIC interpreter sold? Or what if UC Berkeley or Berners-Lee could charge some small fee for each computer running software that uses their work? Surely there is innovation that could have used the funding. But these innovations were successful largely *because* they were free. Any license, fee, or other encumbrance would have hobbled them.

Piracy

Humanity has advanced, when it has advanced,
not because it has been sober, responsible, and cautious,
but because it has been playful, rebellious, and immature.
—Tom Robbins

By the time of Thomas Jefferson's presidency, pirates from the Barbary states of North Africa had inhibited shipping in the Mediterranean Sea for centuries. Many countries such as Great Britain and France paid tribute to keep their ships safe, and American ships benefited from this protection before the Revolutionary War, when they flew the

British flag. Now on its own, the new American government was burdened with paying up to 20 percent of its annual revenue as tribute and ransom. Attacks by a revived American navy in 1805 were the beginning of the end for Barbary piracy.

Pirates have a long history. Plutarch reported pirates in the Mediterranean in the first century BCE. Vikings plundered much of northern Europe during medieval times. English privateers such as Sir Francis Drake captured Spanish treasure ships in the sixteenth century. Closer to home, the North American coast and Caribbean islands were home to many pirates such as Blackbeard and Captain Kidd. "Piracy" today usually means a less violent kind of theft, but it is sobering to remember a time when piracy was done with cutlasses and cannons.

The piracy of intellectual property also has a lengthy history, and copyrights have often been ignored. Nineteenth-century American publishers often pirated British books. While Rudyard Kipling's books sold better in the United States than in England, there were no international copyright laws and Kipling wasn't compensated for his American book sales. Publishers sometimes waited dockside to pick up copies of the latest English titles before their competitors. The reverse was also true. Shortly after its success in the United States, over one million copies of Harriet Beecher Stowe's *Uncle Tom's Cabin* appeared in England without royalties being paid.

Copyright law was initially designed for printed works only, but technology advances pushed it in new directions. In 1893, Thomas Edison had to copyright his first films by printing and copyrighting each individual frame. Photographs themselves weren't copyrightable until 1865, almost forty years after the earliest photographs. Printed sheet music was protected, but that wasn't true for music played through a player piano or phonograph. Sound recordings became copyrightable in 1909. To finally elevate the law above the changing technological details, in 1976 copyrights could be on "any tangible medium of expression, now known or later developed."

Technology has created new avenues for information and, as a result, new avenues for piracy. Software companies have complained about piracy since the first years of the personal computer. Today, the value of pirated or counterfeit software installed on computers worldwide is roughly $30 billion. Software piracy rates are 20 to 30 percent in developed countries but 50 to 90 percent in the developing world.

Worries in the television industry about viewers skipping commercials (what would the advertisers say?) didn't start with VCRs but with the first remote controls, introduced in 1950. A more contentious issue was the ability to record, replay, and even distribute television programs recorded by VCR. The 1984 Supreme Court ruling in the Sony Betamax case said that products that copy are legal if they have "substantial legitimate uses."

This was a landmark intellectual property case, but note that issues with copy machines and libraries preceded it. No one could reprint and sell a copyrighted book without permission, but could a library buy one copy and then let many people read it without paying anything additional to the copyright holder? And is that copyright holder compensated when a book is resold? With the invention of the Xerox copier in 1959, under what conditions could someone make a photocopy of copyrighted material?

Cable and satellite TV have seen a more blatant form of piracy. Video signals on these systems are sometimes scrambled so viewers only see the video if they pay for it. Legitimate set-top boxes would stop working if the bills weren't paid, but clever pirates built set-top boxes that would always work. One pirate argued in court that if the satellite company didn't want him unscrambling their signals, they should stop putting them in his bedroom. Another form of piracy is tapping into cable lines illegally, a utility theft similar to the practice of tapping into electric lines that was common in the early days of the electric industry.

Goods can also be pirated, which takes a slice out of the multibillion-dollar licensing market. A Gucci logo increases the price of a handbag, a Polo logo increases the price of a shirt, and an image of Mickey Mouse increases the price of a toy. With value so out of proportion to the cost of adding the logo or image, the temptation is great. Medicines are also hard to invent but tempting to pirate.

The trait common to all these cases is the large difference between the price charged and the cost of goods. It's very expensive to build the cachet of a brand name or develop a new drug or write a large piece of software or produce a high-budget movie, but to stitch on a Nike logo or produce one more pill of a patented drug or make a copy of Microsoft Windows or *The Matrix* is quite cheap.

Suppose a *Star Wars* fan wonders what *Episode I* would have been like without the Jar Jar Binks character. With some clever editing, the fan makes a new cut of the movie. Is this flattery or parody—or is it plagiarism? And where are the boundaries between an old song, a remixed song, and a new song? Where does borrowing end and piracy begin? All artists borrow from and are inspired by existing work. From the history of the technology-fueled tension between copyright owners and users, we can expect that someone will always be stealing something—and that some of this theft will become part of the next breakthrough.

Though the book and magazine industries feared the photocopier, today they're thriving in spite of it. The movie industry fought against the VCR, though much of their revenue now comes from tape and DVD sales. When consumer digital audio tape (DAT) players were introduced in the early 1990s, the music industry pushed for legislation to cripple their ability to make copies, but the ease of copying CDs and MP3 files soon made that a hollow victory. It has become clear that countries that disregard international copyrights encourage piracy and therefore discourage their own innovators.

The consequences of new technologies are hard to anticipate. A knee-jerk reaction to defend the status quo is understandable, but it is not always the best approach. There is often opportunity in change— and the first company to find it has an important head start.

History has shown that time and market forces
often provide equilibrium in balancing interests,
whether the new technology be a player piano, a copier, a tape recorder,
a video recorder, a personal computer, a karaoke machine or an MP3 player.
—SIDNEY THOMAS, 9th Circuit Court of Appeals judge (2004)

Conclusion:
Vaccinate Against the Hype

One machine can do the work of fifty ordinary men.
No machine can do the work of one extraordinary man.
—ELBERT HUBBARD

In a few minutes a computer can make a mistake so great
that it would have taken many men many months to equal it.
—ANONYMOUS

Logical Fallacies

It's a harsh world out there, with innocent exaggerations and deliberate lies, overenthusiasm and snake oil salesmanship, confusing statistics and insider's jargon, and the human desire to believe and the tendency to distort. To help equip you for the battle, here are a number of fallacies that you should watch out for, both in your own logic and in what you see and hear. Armed with clearer sight, you should be able to better anticipate the future, thereby making more appropriate decisions whether you are looking to acquire technology for a corporation or your personal use.

These fallacies are similar to the High-Tech Myths but are shorter and more focused on errors in logic.

Fast versus Faster Fallacy. We can find many examples of fast change in our lives today, but people in earlier times had their own examples of fast change. To discover if our times are really unique—that our change is fast*er*—we must compare social change today with that in the past.

Avoid being duped by a claim of change without context. Try to find ways to compare that change with change in the past.

Coolness Fallacy. Consider a compact disc—nearly a gigabyte of information on a disc costing pennies. A DVD holds six times more. Or look at the number of transistors that can be put on a semiconductor chip. Or note the complexity of the launching and operation of a communications satellite. We're justifiably impressed by these amazing achievements, but let's see these developments for what they are.

A *technology* might be revolutionary, but the *product* built from that technology won't necessarily be. The operation of a CD, the process of putting more transistors on a chip, and the satellite are technologies; a recording of music, a word processor, and a telephone call might be the corresponding products. A measurement in MIPS, megahertz, or gigabytes (or BTUs, horsepower, or kilowatt-hours) is the measurement of a technology, not a product. Technologies don't have a direct impact—revolutionary or otherwise—on the consumer; the technology *products* have the impact. Keep blinders on to the marvels of the technology and remain objective about the benefits of the product (see also "Avoid Technology Infatuation" on page 28).

Ignorance of Infrastructure Fallacy. Suppose there is a breakthrough in fuel-cell cars a decade from now. We remember that the Web went from 2 percent to 50 percent household penetration in seven years. If the speed at which society adopts technology is getting faster, can we expect even faster adoption for this new fuel cell? Or suppose there is a breakthrough in battery storage technology making electric vehicles competitive alternatives to gasoline cars—will half of new cars be electric just a few years after first rollout? Of course not, in both cases. To ramp up manufacturing, build and staff repair facilities, and create hydrogen or electric filling stations would take much time and money.

Some products need lots of infrastructure (electric or telephone wires, roads, fuel pipelines, or retail outlets, for example) while others don't. Don't sneer at the infrastructure-intensive product by comparing it unfavorably to a quicker-moving product unburdened with infrastructure requirements.

Technologies that have very little need for new infrastructure can be adopted quickly. The Web is an example. Those that require large infrastructure investments have taken much longer and will continue to do so.

Breadth Fallacy. Imagine the launch of a new version of the popular laundry detergent Tide. Most of the existing Tide customers will switch over with their next detergent purchase. That represents growth from zero to tens of millions of customers in just months, much faster than the Internet or any other new technological product has grown. Obviously, as in this example, *breadth* isn't everything. We must also look at *depth*. A new brand of laundry detergent doesn't significantly change the world of many people: it has breadth but little depth. Begging the pardon of Procter & Gamble, consumers would get clothes quite clean with another brand of detergent.

When you look at a new technology, ask yourself if it is like laundry detergent, with breadth but little depth. To properly gauge its impact, we must look at how significantly it changes our lives by considering the way things were done before.

Metcalfe Fallacy. Metcalfe's Law says that the value of a network (an interconnected set of computers or other devices) is proportional to the square of the number of participants. Double the number of possible e-mail recipients or telephone households or fax businesses, and that network becomes *four* times as valuable. But Metcalfe's Law doesn't apply to services. For example, the first Internet bookseller provides a valuable new service. Add another and how much more valuable has the set of booksellers become? Certainly not four times as valuable. Double the number again to four booksellers, and the total value is hardly sixteen times greater than it was with just one. The same is true for search engines, news sites, online encyclopedias, and other services.

We can see this with television. Going from none to one of something is the big jump—the first news program, sitcom, reality show, movie channel, shopping channel, or sports channel. But each new entrant adds less value than the previous one. Increasing selection from ten channels to fifty is not twenty-five times better (as Metcalfe's Law would suggest) or even five times better (if value were simply proportional). Because of redundancy, value increases more slowly. This helps to explain the paradox of one hundred channels of programming and nothing worth watching. With increasing services we must expect diminishing returns. Misapplying Metcalfe's Law is the Metcalfe Fallacy.

Juggernaut Fallacy. DVDs are a better technology than videotapes for movies. Videotapes probably won't coexist with DVDs for long. Similarly, CDs replaced vinyl records, and CD players themselves may be replaced by MP3 players. Yet the newcomer doesn't always supersede the old. Film and radio weren't replaced by television. In fact, they coexist nicely. Similarly, railroads weren't replaced by cars and trucks, nor were conventional ovens by microwave ovens, newspapers by online news services, books by e-books, nor university campuses by online education.

The Internet panicked a number of industries such as higher education and publishing. Instead, each should have been figuring out how the Internet would help them. Avoid this fallacy by realizing that not all technologies are juggernauts, crushing all before them.

Though most haven't been named, other fallacies have already been presented in detail. The following list briefly summarizes some of them and may be a helpful reference.

- *Misdiagnosis Fallacy.* A doctor who incorrectly diagnoses a disease won't prescribe the right treatment. Similarly, we must correctly perceive technology's impact to deal with that impact correctly (see page x).

- *Technological Myopia Fallacy.* Myopia (nearsightedness) shows near things clearly but far things poorly. Technological myopia shows us recent technologies with clarity and emphasis, but older technologies are hazy or ignored. Don't minimize the contributions of past technologies (page 19).

- *The Devil-You-Know Fallacy.* Cars cause two hundred times more annual fatalities than airplanes. Which safety issue deserves more attention? Disease A causes ten times more harm than disease B. Which one should get the greater research focus?

  We are more comfortable with the technology we know than the technology we don't, and harm from this well-known technology is more tolerated. The technology that creates a constant stream of injuries seems more benign than the one that creates the same number of injuries collected into a few disasters. It's also human nature to be more tolerant of natural dangers than technological ones. The result is an unfair comparison of technologies (page 20).

- *Stuck-in-the-Present Fallacy.* The topics of today are often applied too freely into predictions about tomorrow. Even though it's a big deal now, it may not be in the future. Watch out for careless extrapolations of today's trends—none last forever (page 26).

- *Oversimplification Fallacy.* Albert Einstein observed, "Things should be made as simple as possible—but no simpler." Identify and avoid oversimplification. Don't try to squeeze an exponential curve where it doesn't fit or imagine nice patterns when the actual facts are messier (page 63).

- *Wrong Timespan Fallacy.* Don't be so captivated by recent change (timespan #1) that you overlook changes in the past (timespan #2). For example, you might see a huge improvement over the past two hundred years and so credit that mostly to recent inventions. But that isn't always appropriate—sometimes the big jump happened a long time ago. While we have access to more technology now than at any time in history, let's not ignore the substantial foundation from previous generations on which our technology is built. Give the past its due (pages 76, 82, and 119).

- *HDTV Fallacy.* When HDTV finally catches on, will we remember the decades of field trials and missed expectations? No: the press will ignore the disappointing past and marvel instead at how quickly it is changing now. But if we forget the long road that the latest technology traveled, we'll ignore the long road that the *next* new development in the lab will have to endure (page 91).

- *False Novelty Fallacy.* Much has been made about the fast growth of e-commerce, but there are different kinds of electronic purchases. On one hand are purchases that would have been made anyway, even without the Internet. And on the other are purchases that wouldn't have happened without the Internet. Only the latter are new. Most e-commerce sales are *diverted*, not *new*. Be aware of the difference (page 98).

- *Diminishing Returns Fallacy.* Doubling a car's horsepower won't double its top speed. Doubling a pyramid's volume won't double its height. And doubling a PC's speed won't double its usefulness (page 100).

- *Fickleness Fallacy.* "The Atomic Age will provide so many important products! Actually, maybe not so many. But the Space Age will be *really* important. Hmm . . . perhaps that was also oversold. Now, this new Information Age will truly change everything! Well . . ." Watch out for the focus du jour. See past the glossy ads and shiny new toys. Don't jump on the latest bandwagon without good reasons (page 124).

Stay aware of instances of these fallacies as technology is discussed and debated, and look for additional ones. And spread the word.

Technology for the Rest of Us

The hype will continue. Flawed predictions and unjustified euphoria will be common. High-flying tech companies, vetted by well-known and credentialed people, will crash. Claims that paradigms are being broken ("this is unprecedented!") and fundamental axioms must be discarded ("all we need is eyeballs—the money will follow!") will persist.

But you have been vaccinated. You are now more skeptical and better able to see technology accurately. There is nothing special about the rate of new product introduction today or the falling prices of high tech products. The typical product cycle time (the time from innovation to product) is not decreasing. While the availability of trivial information is indeed increasing, fundamental information has always had an outlet. The success of any particular new technology is not inevitable, and change is not exponentially increasing.

The PC and Internet are indeed unprecedented—just like every other major innovation before them. Of course, technology change *does* happen and it can be distressing, but to best handle it we must first see it accurately. Historical examples are a poor fit when shoehorned into an exponential model. The spotlight model is a much better explanation.

I would like to leave you with several suggestions for how to consistently apply this clearer view of technology.

- Don't be bullied into buying a particular technology because a vendor, an advertisement, or your nephew tells you to. Take an active role in what you buy. If it's the right product for you, great, but high tech gets no free ride and must be held accountable, like any purchase.

- Avoid technology infatuation. Cautiously adopt new products, remembering the many failed promises. Users often worry that technology is leaving them behind, but too often the train leaving the station is only an amusement-park ride.

- Resist alarmist claims that technology change is increasing faster and faster, that society is about to be changed beyond recognition, and that you won't be able to handle it. On the contrary, the last two hundred years of technological progress teach us that change is roughly constant, not accelerating; change does indeed happen, but the most extreme predictions are the least accurate; and tomorrow will look more like today than most predictions expect.

In fact, the people who will be most surprised will be those who expect accelerating change.

• Technology is too important to let it take care of itself. We as citizens can't surrender our voice to government and business and assume that they will act in our best interests. Once we have seized that voice, we must use it wisely and demand that technology be applied appropriately. More technology isn't always an improvement.

Taking a more active role in how technology affects your community could mean many things.

Ask how the organizations you're involved with are using or interacting with technology. Is your environmental organization counting on a technology breakthrough to solve a problem? Is your PTA or local government pushing for more computers in the local school without first asking what benefit they will provide (and what that money could be used for instead)? And how do you respond to car ads that boast about horsepower instead of fuel economy?

When you see a shortsighted technological choice made by a government body, don't let it pass. Comment on it, perhaps in a discussion with a co-worker, a letter to the editor, or an e-mail to your representative.

Weigh your personal technology decisions and be aware of whether you're buying for status, for fun, to make yourself more productive, or for other reasons. Any of these could be valid, but understand why you buy. Make your purchase a conscious process in the same way you'd want to be conscious of how advertising affects you.

Get involved in the debate. As an example of a current topic, the United States is no longer the world leader in the use of broadband and cell phones, Internet penetration, or innovation in consumer electronics. Pundits and politicians have pointed to these and other statistics with grave concern. But are these valid indicators of national competitiveness or economic vitality? Maybe, but maybe not. What about the digital divide—the gap between the technology haves and have-nots. Does this still exist? And if it does, does it really matter? This too has been argued both ways. What do you think about where our energy comes from, about global warming, or about how much the federal government spends on space research? Dozens of similar issues are being debated right now. If this book has been successful, you should feel better prepared to deal with the perpetual technology revolution and separate the reality from the hype.

Philosophical habits of mind do not come quicker through fiber optics.
Clear thinking is not aided by better dot resolution.
Understanding ourselves and feeling for others
does not come with a software upgrade.

—LINDA RAY PRATT, president,
American Association of University Professors (1994)

Notes

Preface

xi   *cost of information technology and amount wasted:* "Worldwide, companies waste as much as 20% of the $2.7 trillion spent annually on tech." Jim Hopkins and Michelle Kessler, "Companies Squander Billions on Tech," *USA Today,* May 20, 2002, www.usatoday.com/usatonline/20020520/4124243s.htm.

xi   *public interest in science:* National Science Board, "Science and Engineering Indicators 2002," www.nsf.gov/sbe/srs/seind02; "Stories of the Century: The Nation Votes," *USA WEEKEND* and Newseum, 1999, www.newseum.org/century/finalresults.htm.

Introduction: Leveling the Exponential Curve

1   *exponential increase in grains of rice:* there are roughly 12,000 grains of rice (weighing 6.75 ounces) in a cup of rice. The rest is an exercise to the reader.

1   *Mach 700:* A top speed of 131 mph $\times 2^{12} = 700 \times$ Mach 1 (765.6 mph at sea level).

4   *falling populations in many countries:* The UN reports that sixty-one countries now have below-replacement fertility. See www.unpopulation.org.

1. The Birthday-Present Syndrome

11   *malaria incidence in India:* Melvin Kranzberg, "Technology and History: Kranzberg's Laws," *Technology and Culture,* July 1986, 546.

11 *DDT still used widely:* Darwin Stapleton, "The Short-Lived Miracle of DDT," *Invention & Technology,* Winter 2000, 34.

11 *Kranzberg's First Law:* Kranzberg, "Technology and History," 545.

11 *Every law creates an outlaw:* John Heider, *The Tao of Leadership* (New York: Bantam, 1985), chapter 57.

12 *avoiding certain areas of research:* Bill Joy, "Why the Future Doesn't Need Us," *Wired,* April 2000, www.wired.com/wired/archive/8.04/joy.html.

13 *original paper about computer chess:* Claude Shannon, "A Chess-Playing Machine," *Scientific American,* February 1950, 48–51.

14 *typesetting competitions:* Walker Rumble, "Ready, Go, SET!" *Invention & Technology,* Spring 2001, 40.

16 *the Turk: Popular Mechanics,* "Talking Robots," June 1, 2000, www.popularmechanics.com/science/time_machine/1288696.html.

2. The Perils of Prediction

22 *"drop in by rocket plane":* Waldemar Kaempffert, "Miracles You'll See in the Next 50 Years," *Popular Mechanics,* February 1950, 112.

22 *only a quarter of predictions were accurate:* Steven Schnaars, *Megamistakes* (New York: The Free Press, 1989), 32.

23 *all trees gone by 1920:* Richard Lacayo, "Future Schlock," *Time,* Fall 1992, 90.

23 *fast electric ships:* John E. Watkins, "What May Happen in the Next Hundred Years," *The Futurist,* October 1982.

23 *smiling Garden of Eden:* this was said by Frederick Soddy, one of the scientists who discovered radioactivity. See Brian Horrigan, "Grand Visions: The History of Predicting the Future," *Encarta Yearbook,* August 1999.

23 *animal parts for food:* one predictor of this was Winston Churchill. See Helen O'Neill, "Future Didn't Quite Live Up to All the Fanfare," *Chicago Daily Herald,* www.dailyherald.com/special/crossingcenturies/2a/4rest.asp.

23 *dome over city:* J. Baldwin, *Bucky Works* (New York: Wiley, 1997), chapter 9.

24 *moon bases and robot soldiers:* Schnaars, *Megamistakes,* 10.

24 *electromagnetic fields for classrooms:* William Shanklin, *Six Timeless Marketing Blunders* (New York: Lexington, 1988), 77.

25 *falcons at JFK:* Mary Pflum, "Earth Matters: Bird-Plane Collisions on the Rise," *CNN,* November 24, 2000, http://archives.cnn.com/2000/NATURE/11/24/birds.planes/.

27 *workers displaced by robots:* Merrill Sheils et al., "And Man Created the Chip," *Newsweek,* June 30, 1980, 55.

27 *farming in the future:* Schnaars, *Megamistakes,* 13.

27 *railroad and pneumatic tubes:* Peter Denning, *Talking Back to the Machine* (Springer, 1999).

28 *"In the space of 176 years":* Mark Twain, *Life on the Mississippi,* chapter 17 (originally published in 1883).

29 *the future home:* Eric Lefcowitz, "21st Century Space-Age Bachelor Pad," Retrofuture. http://retrofuture.com/spaceage.html.

29 *paper dresses:* "Products of Yesteryear," Kimberley-Clark, www.kimberly -clark.com/aboutus/paper_dresses.asp.

29 *fax newspapers:* George Mannes, "Delivering the FAX," *Invention & Technology,* Spring 1999, 40–48.

33 *VHS versus Beta:* Mark Levy, *The VCR Age* (London: Sage Publications, 1989), 28; S. J. Liebowitz and Stephen E. Margolis, "Market Processes and the Selection of Standards," wwwpub.utdallas.edu/~liebowit/standard/standard.html.

33 *QWERTY versus Dvorak:* S. J. Liebowitz and Stephen E. Margolis, "Typing Errors," *Reason Online,* June 1996, www.reason.com/9606/Fe.QWERTY .html.

3. The Unintended Wager

37 *DDT in Borneo:* F. Y. Cheng, "Deterioration of thatch roofs by moth larvae after house spraying in the course of a malaria eradication programme in North Borneo," *Bulletin of the World Health Organization* 28(1) (1962):136–37.

38 *hygiene hypothesis:* Karen Wright, "Gut Reaction," *Discover,* February 2003, 4.

39 *hospital infections:* Centers for Disease Control and Prevention, "Public health focus: surveillance, prevention, and control of nosocomial infections," *Morbidity and Mortality Weekly Report,* October 23, 1992, www .cdc.gov/mmwr/preview/mmwrhtml/00017800.htm.

39   *Cluetrain Manifesto:* Chris Locke, Doc Searls, and David Weinberger, "The Longing," *The Cluetrain Manifesto,* 1999, www.cluetrain.com/book/index.html.

40   *e-mail chain letters:* One good site for debunking information is www.snopes.com.

40   *get-well cards for sick boy:* David Emery, "A user's guide to Craig Shergold," http://urbanlegends.about.com/library/weekly/aa102997.htm.

41   *the Web and the bandwagon effect in science:* James Glanz, "The Web as Dictator of Scientific Fashion," *New York Times,* June 19, 2001, www.nytimes.com/2001/06/19/science/19NET.html.

41   *unfounded fears in the entertainment industry:* John Perry Barlow, "The Next Economy of Ideas," *Wired,* October 2000, 240; David McGuire, "Study: File-Sharing No Threat to Music Sales," *Washington Post,* March 29, 2004, www.washingtonpost.com/ac2/wp-dyn/A34300-2004Mar29.

41   *volume of paper use and paperless bathrooms:* Al Ries and Jack Trout, *Positioning* (New York: Warner Books, 1986), 12; "Paperless Office? Don't Wait!" *MIS Week,* June 30, 1986, 40.

42   *household appliances:* Ruth Cowan, "Less Work for Mother?" *Invention & Technology,* Spring 1987, 57.

43   *power consumed when appliances off:* Rachael Moeller Gorman, "Wasting Away on Standby," *Discover,* December 2002, 13.

43   *comparative costs of dumping and recycling:* Gretel H. Schueller, "Wasting Away," SF Environment, Fall 2002, www.sfenvironment.com/articles_pr/2002/article/090002.htm.

43   *consequences of freon:* Mark Bernstein, "Thomas Midgley and the Law of Unintended Consequences," *Invention & Technology,* Spring 2002, 38.

45   *"You are now a Systems-person":* John Gall, *Systemantics* (Ann Arbor, MI: The General Systemantics Press, 1986), 129.

46   *wicked versus tame problems:* Rittel and Webber, "Dilemmas in a General Theory of Planning," *Policy Sciences,* 4 (1973): 155–69.

4. If It Ain't Broke, Be Grateful

49   *burning of library at Alexandria:* Preston Chesser, "The Burning of the Library of Alexandria," eHistory.com, www.ehistory.com/world/articles/ArticleView.cfm?AID=9.

50 *multimedia version of Doomsday Book*: Robin McKie and Vanessa Thorpe, "Digital Domesday Book Lasts 15 Years not 1000," *Guardian Unlimited,* March 3, 2002, http://books.guardian.co.uk/news/articles/0,6109,661585,00.html.

51 *"Almost everything today gets recorded":* Jack Hitt, "How to Make a Time Capsule," *New York Times Magazine,* December 5, 1999, www.nytimes.com/library/magazine/millennium/m6/capsule-hitt.html.

52 *lost U.S. census and NASA data:* Emma Cobb, "Where Have All Our Records Gone?" *Invention & Technology,* Fall 1986, 8; David Propson, "NASA's Backup Backup," *Wired,* November 2000, 114, www.wired.com/wired/archive/8.11/mustread.html?pg=9.

52 *obsolete media and drives:* Emma Cobb, "Where Have All Our Records Gone?" *Invention & Technology,* Fall 1986, 8–9; Edward Tenner, "We the Innovators," *US News and World Report,* January 3, 2000, www.usnews.com/usnews/culture/articles/000103/archive_033967.htm.

53 *losing Web history and bulletin board systems:* Nick Montfort, "In Search of Webs Past," *Technology Review,* July 2000, 105; Katharine Mieszkowski, "Relics of the Lost Bulletin-Board Tribes," *Salon.com,* January 22, 2002, www.salon.com/tech/feature/2002/01/22/bbs_archives/print.html.

53 *temporary photos at Web sites:* Tenner, "We the Innovators," *US News and World Report.*

53 *problems of preserving old software:* Daniel Terdiman, "Fighting to Preserve Old Programs," *Wired,* October 14, 2003, www.wired.com/news/culture/0,1284,60770,00.html.

54 *proving prior art in patent battles:* Claire Tristram, "Preserve Your Data Forever," *Technology Review,* October 2002, 38.

54 *bank credit and Mars Climate Orbiter bugs:* Rich Hower, "What Are Some Recent Major Computer System Failures Caused by Software Bugs?" Software QA/Test Resource Center, www.softwareqatest.com/qatfaq1.html#FAQ1_3; Isbell, Hardin, and Underwood, "Likely Cause of Orbiter Loss Found," http://mars.jpl.nasa.gov/msp98/orbiter.

55 *upgrade to the air traffic control system:* the program is called STARS (Standard Terminal Automation Replacement System).

55 *cost of software bugs:* Research Triangle Institute (project number 7007.011), "The Economic Impacts of Inadequate Infrastructure for Soft-

ware Testing," May 2002, ES-1 and 8-1, www.mel.nist.gov/msid/sima/
sw_testing_rpt.pdf.

56 *cost of spam, viruses, and spyware:* for Brightmail e-mail analysis, see
www.brightmail.com; The Center for Excellence in Service, University of
Maryland, "2004 National Technology Readiness Survey," February 3,
2005, www.rhsmith.umd.edu/ntrs/NTRS_2004.pdf; 2005 *Consumer
Reports* State of the Net survey, September 2005.

56 *burglar alarms:* Edward Tenner, *Why Things Bite Back* (New York: Knopf,
1996), 230.

58 *paper plant in Amazon:* Bill Adler and Julie Houghton, *America's Stupid-
est Business Decisions* (New York: Quill, 1997), 189.

61 *cost of laptop theft:* Richard Power, "2002 CSI/FBI Computer Crime and
Security Survey," *Computer Security Issues & Trends,* Spring 2002 (Computer Security Institute), 10–11, http://i.cmpnet.com/gocsi/db_area/
pdfs/fbi/FBI2002.pdf.

62 *military and other hackers:* Joanna Glasner, "Y2.02K: Future Schlock?"
Wired, September 2, 1999, www.wired.com/news/culture/0,1284,21553,00
.html.

62 *solar storms:* James Burch, "The Fury of Space Storms," *Scientific American,* April 2001, 86.

5. More Powerful Than a Locomotive

64 *"The world of today":* Alvin Toffler, *Future Shock* (New York: Bantam,
1970), 13.

64 *Law of Accelerating Returns:* Ray Kurzweil, *The Age of Spiritual Machines*
(New York: Penguin, 1999), 30.

68 *"I have seen the future":* James Gleick, *What Just Happened* (New York:
Pantheon, 2002), 9.

69 *chart of airspeeds:* Robert Zubrin and Mitchell Clapp, "Aviation's Next
Great Leap," *Technology Review* January 1998, 30, www.technology
review.com/articles/zubrin0198.asp.

69 *production figures for small planes:* "Table 1–12: Sales or Deliveries of
New Aircraft, Vehicles, Vessels, and Other Conveyances," *National Transportation Statistics, 2003,* Bureau of Transportation Statistics, www.bts
.gov/publications/national_transportation_statistics/2003/html/table
_01_12.html.

69 *battery* vs. *sugar:* An alkaline battery has 587 j/g of energy, and sugar has 4120 j/g. There are about three calories of energy in one AA alkaline battery.

70 *problems with electric cars:* How Stuff Works, "How Electric Cars Work," http://auto.www.howstuffworks.com/electric-car4.htm.

70 *"Never in modern history":* Michio Kaku, "Nuclear Power," *The Reader's Companion to American History,* http://college.hmco.com/history/readerscomp/rcah/html/ah_065600_nuclearpower.htm.

72 *"I have a rule":* Andrew Grove, "What Can be Done, Will be Done," *Forbes ASAP,* December 2, 1996.

72 *"In accordance with the Law of Accelerating Returns":* Kurzweil, *The Age of Spiritual Machines,* 33, 253.

74 *"Find a block which is taller":* Terry Winograd's Stanford homepage, http://hci.stanford.edu/winograd/.

74 *"have the most sweeping implications":* "Technologies for the '80s," *Business Week,* July 6, 1981, 50.

75 *"Over the next several decades":* Kurzweil, *The Age of Spiritual Machines,* 4.

77 *the Battle of New Orleans:* David Eggenberger, *An Encyclopedia of Battles* (New York: Dover, 1985), 303.

6. Faster Than a Speeding Bullet

80 *"new inventions now arrive at a bewildering rate":* "Social Time," *Forbes ASAP,* November 30, 1998, www.forbes.com/asap/98/1130/253.htm.

80 *"More than half of U.S. patents":* W. Michael Cox and Richard Alm, "The Economy at Light Speed," *1996 Annual Report* (Federal Reserve Bank of Dallas, 1996), 7, www.dallasfed.org/htm/pubs/annual.html.

81 *number of new products introduced:* U.S. Census Bureau, "New Product Introductions of Consumer Packaged Goods: 1980 to 1997," *Statistical Abstract of the United States: 1998,* Table 889, 558.

81 *rate of patent application:* Robert Adler, "Entering a Dark Age of Innovation," *New Scientist,* July 2, 2005, www.newscientist.com/article/mg 18625066.500.

81 *1920s brand longevity:* Philip Kotler, *Marketing Management, the Millennium Edition* (Prentice Hall, 1999), 307.

82 *"A weekday edition of the New York Times":* Richard Saul Wurman, *Information Anxiety* (New York: Doubleday, 1989), 32.

82 *"The total of all printed knowledge"*: Peter Large, *The Micro Revolution Revisited* (London: Frances Pinter, 1984), 46.

84 *changing cost of consumer goods:* W. Michael Cox and Richard Alm, "Time Well Spent: The Declining *Real* Cost of Living in America," *1997 Annual Report* (Federal Reserve Bank of Dallas, 1997), www.dallasfed .org/htm/pubs/annual.html.

85 *"Radio was in existence 38 years"*: "The Emerging Digital Economy," U.S. Department of Commerce, April 1998, 4, www.esa.doc.gov/TheEmerging DigitalEconomy.cfm.

89 *technologies that proceeded the Internet:* Carol Wilson, "The Myths and Magic of Minitel," *Telephony,* December 2, 1991, 52; Larry Yokell, "Information Services: From Start to Present Day," *Specs* (CableLabs, Inc.), 1992.

90 *increased press coverage of crime:* Barry Glassner, *The Culture of Fear* (New York: Basic Books, 1999) xi–xxi.

91 *history of HDTV:* Corey P. Carbonara, "A Current History of High Definition Television," *HDTV World Review,* Autumn 1990, www.hdtvmagazine .com/archives/wrldrvw3.html.

7. Leap Tall Buildings in a Single Bound

94 *newspapers for everyone:* "Electronic Newspaper Is an Oxymoron," *Digital Media,* May 10, 1995, 14.

96 *size of Internet industry:* Eric Nee, "The Net Really is Big," *Fortune.com,* June 11, 1999. See also "The Internet Economy Indicators," Key Findings, www.internetindicators.com/keyfindings.html.

97 *the real size of the Internet economy:* "The Internet Economy Indicators," Key Findings, www.internetindicators.com/keyfindings.html.

97 *e-commerce sales:* "Retail Trade Sales," *Statistical Abstract of the United States 2002,* Table 1021, www.census.gov/prod/www/statistical-abstract -02.html.

99 *"the Internet changes everything"*: J. Neil Weintraut, quoted in "The Software Revolution—Part 1," *Business Week,* December 4, 1995.

100 *relationship between exponential growth and demands:* Phil Agre, "Some notes about distributed objects, technology-driven change, and the diversity of knowledge," January 27, 2002, http://commons.somewhere .com/rre/2002/RRE.notes.and.recommenda.html.

101 *success of containerized shipping:* Peter Drucker, *Innovation and Entrepreneurship* (New York: HarperBusiness, 1985), 63.

102 *Moore's Second Law:* Michael Malone, "Forget Moore's Law," *Red Herring*, February 10, 2003.

103 *technologies in schools:* Todd Oppenheimer, "The Computer Delusion," *The Atlantic*, July 1997, www.theatlantic.com/issues/97jul/computer.htm.

103 *assessment of technology and public education:* Congressional Office of Technology Assessment, *Teachers and Technology: Making the Connection*, OTA-HER-616 (GPO stock #052-003-01409-2), April 1995, 104, www.wws.princeton.edu/~ota/ns20/alpha_f.html.

104 *the fortunes of The Learning Company:* Matt Richtel, "Once a Booming Market, Educational Software for the PC Takes a Nose Dive," *New York Times*, August 22, 2005, www.nytimes.com/2005/08/22/technology/22soft.html.

104 *"These students who have less access":* Kevin Finneran, "Let Them Eat Pixels," *Issues in Science and Technology*, Spring 2000, http://search.nap.edu/issues/16.3/editorsjournal.htm.

106 *expenditures on computers versus returns:* Hopkins and Kessler, "Companies Squander Billions on Tech," *USA TODAY*, May 20, 2002, www.usatoday.com/usatonline/20020520/4124243s.htm.

107 *billions wasted on Internet surfing and computers:* A 2005 survey estimated that the average U.S. worker wastes more than an hour more per day than expected, which adds up to almost $800 billion per year. Close to half of that is due to personal Internet surfing. See "Wasted Time at Work Costing Companies Billions" at www.salary.com; W. Wayt Gibbs, "Taking Computers to Task," *Scientific American*, July 1997, 87.

107 *"You see computers everywhere":* Robert Solow, "We'd Better Watch Out," *New York Times Book Review*, July 12, 1987, 36.

107 *business needs to learn to use computers:* Kristin Leutwyler, "Productivity Lost," *Scientific American*, November 1994, 101.

8. Corrective Lenses

111 *relative importance of computers:* Bill Gates made this point at the Digital Dividends Conference, October 18, 2000. See www.microsoft.com/billgates/speeches/2000/10-18digitaldividends.asp.

114   *the Cambrian Explosion:* Madeleine Nash, "When Life Exploded," *Time,* December 4, 1995, 66.

114   *the basics have already been invented:* Georges Polti, *The Thirty-Six Dramatic Situations* (Franklin, Ohio: James Knapp Reeve, 1921); Levy and Salvadori, *Why Buildings Fall Down* (New York: Norton, 1987), 287.

117   *estimations of extinct species:* "Mass Extinction," American Museum of Natural History, www.amnh.org/exhibitions/dinosaurs/extinction/mass .php.

9. For Better or For Worse

124   *"Today no prudent businessman"*: "What is the Atom's Industrial Future?" *Business Week,* March 8, 1947, 21-2.

124   *"unlimited power"*: Lewis L. Strauss, Chairman, U.S. Atomic Energy Commission, "Remarks Prepared for Delivery at the Founders' Day Dinner, National Association of Science Writers" September 16, 1954.

124   *Project Plowshare:* Robert Pool, *Beyond Engineering* (New York: Oxford University Press, 1997), 71.

126   *the appeal of appending ".com"*: Cooper, Dimitrov, and Rau, "A Rose.com by Any Other Name," *The Journal of Finance,* December 2001, 2371, www .mgmt.purdue.edu/faculty/mcooper/FinalJFversion2371-2388.pdf; Lorrie Grant, "What's in a Name? The Fading of Dot-com," *USA TODAY,* July 3, 2000, 1A.

130   *the threat television poses to other cultures:* Mark Kurlansky, *The Basque History of the World* (Walker, 1999); Scott-Clark and Levy, "Fast Forward into Trouble," *The Guardian,* June 14, 2003, www.guardian.co.uk/Print/ 0,3858,4689384,00.html.

131   *self-portrait by Paula Scher:* Alan Fletcher, *The Art of Looking Sideways* (London: Phaidon, 2001), 341.

131   *"by the sane and sensible citizen"*: Claude Fischer, *America Calling* (Berkeley, CA: University of California Press, 1992), 225.

132   *Max Nordau's predictions:* Stephen Kern, *The Culture of Time and Space, 1880–1918* (Cambridge, MA: Harvard University Press, 1983), 70.

132   *article in Smithsonian about sleep disruption:* Joyce and Richard Wolkomir, "When Bandogs Howle & Spirits Walk," *Smithsonian,* January 2001, 39.

132   *sleep patterns of American adults:* James Gleick, *Faster: The Acceleration of Just About Everything* (New York: Pantheon, 1999), 121.

134   *wristwatch for men:* Charles Panati, *Panati's Parade of Fads, Follies, and Manias* (New York: HarperPerennial, 1991), 102.

136   *telephone joke:* Mark Caldwell, *A Short History of Rudeness* (New York: Picador, 1999), 131.

137   *German telephone conference:* Peter Drucker, "Infoliteracy," *Forbes ASAP,* August 29, 1994, 105.

137   *typewritten letters:* Cynthia Monaco, "The Difficult Birth of the Typewriter," *Invention & Technology,* Spring/Summer 1988, 11.

138   *machine guns and heroism:* Kirkpatrick Sale, *Rebels Against the Future* (Reading, MA: Addison-Wesley, 1995) 17.

10. Playing with Matches

139   *major floods in China in 1975 and 1931:* "Banqiao Dam," *Wikipedia,* http://en.wikipedia.org/wiki/Banqiao_Dam; "Huang He," *Wikipedia,* http://en.wikipedia.org/wiki/Huang_He; and "1931 Huang He flood," *Wikipedia,* http://en.wikipedia.org/wiki/1931_Huang_He_flood.

142   *the demise of the passenger pigeon:* Emma Cobb, "Victim of Technology," *Invention & Technology,* Spring 1987, 6–7.

142   *villages lost in England during the plague:* "Black Death," *Encyclopaedia Britannica,* www.britannica.com/eb/article-9015473.

143   *the age of Juliet:* Michio Kaku, *Visions* (New York: Anchor, 1998), 203.

143   *beer more pure than water:* Bert Vallee, "Alcohol in the Western World," *Scientific American,* June 1998, 81.

145   *Mohawk ironworkers:* David Monthorn, "Mohawk Ironworkers Build New York," *News from Indian Country,* www.indiancountrynews.com/ironworkers.cfm.

147   *"Is this what your wife should be cooking with?":* John Lienhard, *The Engines of Our Ingenuity* (New York: Oxford University Press, 2000), 151.

147   *deaths due to kerosene:* Ron Chernow, *Titan* (New York: Random House, 1998), 253.

148   *Bulletin of the Atomic Scientists:* www.thebulletin.org.

148   *noxious manure:* Justin Hyde, "The Creation of the Automotive World," *The*

WIRE, November 28, 1999, http://wire.ap.org/APpackages/20thcentury/
centstories/112899cars.html.

150 *"You'll just be poor.":* story related by Dr. William Foege at the World
Health Assembly 2000, www.gatesfoundation.org/connectedpostings/
hmaga_sn.htm.

150 *Iridium satellite phones:* Sam Silverstein, "Iridium Looks to Commercial
Business for Continued Success," *Space News Business Report,* October
20, 2003, www.space.com/spacenews/archive03/iridiumarch_102003
.html.

151 *risk assessment versus real danger:* Steve Mirsky, "Dumb, Dumb, Duh
Dumb," *Scientific American,* November 2001, 95.

11. Fear and Anxiety

155 *"powers of darkness":* Richard Scheffel, ed. *Discovering America's Past*
(Pleasantville, NY: Reader's Digest, 1993), 360.

158 *images of Balinese witches:* Pico Iyer, *Video Night in Kathmandu* (New
York: Knopf, 1988), 53.

158 *"The steady evaporation of the question":* Denis Dutton, "When Will Overpopulation Create Worldwide Starvation?" *The Edge,* 2001, www.edge
.org/documents/questions/q2001.html.

159 *consider the number four unlucky:* The Las Vegas hotel without floors 40–
49 is Rio Suites.

161 *how Pacific islanders saw galleons:* Lawrence Weschler, *Mr. Wilson's Cabinet of Wonder* (New York: Vintage 1995), 120.

161 *cargo cults:* Richard Feynman, "Cargo Cult Science," Caltech commence-
ment address 1974, www.physics.brocku.ca/etc/cargo_cult_science.html.

161 *telephone stage fright:* Robert Lucky, *Silicon Dreams* (New York: St. Martin's Press, 1989), 202.

165 *need for switchboard operators:* Peter Drucker, *Innovation and Entrepreneurship* (New York: HarperBusiness, 1985), 69.

12. Technologies That Touch Us

167 *biomorphic designs:* Doug Stewart, "Cheese Holes, Blobs, and Woggles,"
Smithsonian, February, 2002, 40.

167 *rocket-inspired cars:* Lienhard, *The Engines of Our Ingenuity,* 16.

168   *new products that used kerosene:* Chernow, *Titan*, 253.

168   *"It is astonishing how, in a few years":* Lienhard, *The Engines of Our Inge-nuity*, 233.

169   *polyurethane houses:* Paul Kirchner, *Forgotten Fads and Fabulous Flops* (Los Angeles, CA: Rhino, 1995), 88.

170   *failure of the picturephone:* Lucky, *Silicon Dreams*, 301.

171   *exponential clock improvement:* Lienhard, *The Engines of Our Ingenuity*, 129.

171   *Harrison's accurate chronometers:* Ola and d'Aulaire, "Taking the Mea-sure of Time," *Smithsonian*, December 1999, 53.

172   *"penetrate so deeply, so tyrannically":* Stephen Kern, *The Culture of Time and Space, 1880–1918* (Cambridge: Harvard University Press, 1983), 125.

172   *crossing time zones:* Ibid., 11.

174   *Internet Time:* "Swatch Internet Time," *Wikipedia,* http://en.wikipedia.org/wiki/Swatch_Internet_Time.

176   *"The delicacy, intricacy, and nuance of language":* Kern, *The Culture of Time and Space*, 115.

180   *Telefon:* David Crystal, *The Cambridge Encyclopedia of Language* (Cam-bridge, UK: Cambridge University Press, 1997), 4.

183   *binary prefixes:* "Prefixes for Binary Multiples," International Electro-technical Commission, http://physics.nist.gov/cuu/Units/binary.html.

13. Innovation Stimulation

185   *Roman slave labor:* David Lance Goines, "An Apologia for Anarchism," 1985, www.goines.net/Writing/apologia_for_anarc.html.

186   *.com adds value:* Mary Kwak, "A Rose.com by Any Other Name," *Inc,* June 2000, www.inc.com/articles/2000/06/19489.html.

189   *stock values of the company du jour:* Justin Fox, "How New is the Inter-net, Really?" *Fortune*, November 22, 1999; William Holstein, "To Gauge the Internet, Listen to the Steam Engine," *New York Times*, August 26, 2001, www.nytimes.com/2001/08/26/business/26SVAL.html.

191   *Feynman's nanotechnology prize:* Richard Feynman, "There's Plenty of Room at the Bottom," 1959, www.zyvex.com/nanotech/feynman.html; see also www.foresight.org/GrandPrize.1.html.

191   *other to-be-won prizes:* see www.arpa.gov/grandchallenge/overview.html;
       see www.eff.org/awards/coop.php; see http://loebner.net/Prizef/loebner
       -prize.html; see www.longbets.org.

192   *Lindbergh's U.S. tour:* The Straight Dope, "Did others fly across the Atlan-
       tic before Lindbergh?" March 25, 2003, www.straightdope.com/mailbag/
       mtransatlantic.html.

193   *dispelling legends:* Daniel J. Boorstin, *The Discoverers* (New York: Vintage
       Books, 1983), 106.

194   *Marco Polo's trip through Asia:* "Marco Polo's Guide to Afghanistan,"
       Smithsonian, January 2002.

194   *Japanese soldiers finally surrender:* "Japanese Holdouts in the Pacific,"
       www.wanpela.com/holdouts/list.html.

194   *the era of clipper ships:* Nicholas Dean, "The Brief, Swift Reign of the Clip-
       pers," *Invention & Technology,* Fall 1989, 48.

196   *"England had gone in a generation":* paraphrased from Friedrich Engels,
       The Conditions of the Working-Class in England (1845), "The Great Towns,"
       www.marxists.org/archive/marx/works/download/Engels_condition_of
       _the_Working_Class_in_England.pdf.

198   *"[Beneath a bridge] flows, or rather stagnates":* Engels, *The Conditions of
       the Working-Class in England.*

199   *percentage of population in manufacturing:* Asa Briggs and Daniel Snow-
       man, eds., *Fins de Siècle: How Centuries End 1400–2000* (New Haven, CT:
       Yale University Press, 1996), 129.

14. What's Mine Is Mine

201   *the truth about the patent commissioner's statement:* John Horgan, "The
       Twilight of Science," *Technology Review,* July 1996, 57.

201   *loss of music sales?* David McGuire, "Study: File-Sharing No Threat to
       Music Sales," *Washington Post,* March 29, 2004, www.washingtonpost
       .com/ac2/wp-dyn/A34300-2004Mar29.

202   *Byzantine monopoly on silkworms:* Philip Kotler, *Marketing Manage-
       ment, the Millennium Edition* (Upper Saddle River, NJ: Prentice Hall,
       1999), 397.

203   *history of cochineal:* Françoise Delamare and Bernard Guineau, *Colors*
       (New York: Abrams, 1999), 70, 133.

204 *the spread of textile factories:* Lienhard, *The Engines of Our Ingenuity*, 113.

205 *Bell versus Gray: Inventor's Digest,* July/August 1998, 26–28.

205 *Alexander Bain and his automatic telegraph:* Tim Hunkin, "Just Give Me the Fax," *New Scientist,* February 13, 1993, 33.

207 *"I considered the whole idea"*: Frederic Schwarz, "The Patriarch of Pong," *Invention & Technology,* Fall 1990, 64.

208 *"no inventor has the right to profit"*: John O'Rourke, "Among the Works of God and Man," *Invention & Technology,* Spring/Summer 1998, 64.

210 *value of pirated software: SiliconValley.com,* "36 Percent of Software Worldwide Pirated, Trade Group Says," July 7, 2004, www.siliconvalley .com/mld/siliconvalley/9097724.htm.

Conclusion: Vaccinate Against the Hype

220 *no longer the leader:* "A Nation Online: How Americans Are Expanding Their Use of the Internet," U.S. Department of Commerce, February 2002, www.esa.doc.gov/nationonline.cfm; "High-Speeders Increase, But . . ." CBS News, November 23, 2004, www.cbsnews.com/stories/2004/11/23/ tech/main657246.shtml.

Acknowledgments

Ruffling a few feathers was a goal of this book, but I'm sure there are places where I overreach, my argument needs more support, or my point is unclear. If not for the help of many people, the number of those errors would be far greater.

Thanks are due to Steve Piersanti, who had faith in me and saw in the manuscript the book that it could become; Jeevan Sivasubramaniam, who was always ready with wise counsel; and the great Berrett-Koehler staff, who saw this project to completion. Thanks are also due the many other B-K authors, who made me feel welcome. Chief among these was Gifford Pinchot, who made the introduction to Steve and started this process.

Many sharp eyes combed the manuscript for errors: Valerie Andrews, Jeff Kulick, Jennifer Liss, Eric Lopez, Katherine Silver, and Joseph Webb, plus Pamela Greenwood and Heidi Wrightsman from Seattle's Author-Editor Clinic. I also received helpful comments and encouragement from Mark Anderson, Jesse Berst, Ray diCasparro, Cris DiMarco, Jennifer DiMarco, Chris Doerr, Neil Gerrans, David Green, Liz Seidensticker, Verna Seidensticker, Rick Townsend, and Jack Turk.

To the technology statesmen who wrote endorsements for this book, lending me a bit of your well-earned stature, thank you. And thank you Microsoft, the velvet sweatshop, for hosting an eight-year high-tech boot camp for my personal edification.

The many friends I've talked to about the book have helped me hone my "elevator pitch." Whether their reaction was agreement or argument, it was quite valuable. My family has been supportive from the beginning, graciously putting up with my seven-year affair with this project.

And finally, The Wednesday Night Chocolate Society has for more than two years read my stuff, pointed out its flaws, helped me write better, and cajoled me to finish this book. "If it was hard to write, it should be hard to read" wouldn't do. Thanks Brian Green, Gilla Bachellerie, Chris McFaul, George Boyle, Donald Gilbert-Santamaria, Kim Ritchie, and Donna Fitzgerald.

Index

About the Author

I learned how to program in high school in the mid-1970s on a computer designed in 1962. It had four cubes of core memory each the size of a coffee cup (holding roughly 64K bytes) and two 10-megabyte disk drives each bigger than a car tire. Teletype terminals in the school connected to the computer with a 110 bit-per-second telephone modem. From that point to the present, I've taken a ringside seat and watched technology change with fascination.

After graduating from MIT with a degree in Computer Science and Engineering, I designed digital hardware, about which I wrote my first book, *The Well-Tempered Digital Design* (1986). I have also programmed in a dozen computer languages and in environments ranging from punch cards to one of the first windowing environments, to MS-DOS, to Windows (starting with version 1.0). Along the way, I picked up thirteen software patents.

The computer industry is enormously varied, and I've had the privilege of seeing it from within a number of companies. The biggest was IBM; the smallest was Television Laboratories, a startup of about ten people. In 1986, back in the PC dark ages, our engineering team designed and built a high-end PC video card with three custom gate arrays running at 100 MIPS and driven by video paint and presentation software—basically PowerPoint a decade ahead of time. My primary contribution was about seventy thousand lines of code.

When I began working at Microsoft in 1989, it was already a large company. Nevertheless, we joked that not only was Microsoft not the biggest software company in the world, it wasn't even the biggest software company in *Redmond* (it competed for that title with Nintendo's U.S. headquarters). During my eight years there, it increased tenfold to about forty thousand employees.

At Microsoft, I was a program manager on a number of projects, including Windows 3.0 (no, I wasn't the only one) and the first ROM-based version of Windows. The project I'm most proud of was Nomad, a tiny operating system designed for pagers, watches, and similar small devices.

Since leaving Microsoft, this book has been my major project. I've also dabbled with an education software company and third-world travel, including memorable visits to Peru, Malawi, and Mongolia.

The genesis of this book was at Microsoft, where I began researching technology change. I discovered that much of technology's long history has been distilled to where it resembles a myth—something like, "And lo, Betamax was created from nothing. VHS followed soon after. And there was a great battle, with VHS reigning supreme." The actual history turned out to be more involved—and a lot more interesting—than these summaries. More important, the popular, simplified perception of technology change was off the mark. An important story about how technology change really happens wasn't being told. That is the story in this book.

If you want to see the latest on this topic, go to the book's Web site, www.future-hype.com. And if you have comments about this book or technology's impact on society, I'd like to hear them. Contact me at bob@future-hype.com.

About Berrett-Koehler Publishers

Berrett-Koehler is an independent publisher dedicated to an ambitious mission: Creating a World that Works for All.

We believe that to truly create a better world, action is needed at all levels—individual, organizational, and societal. At the individual level, our publications help people align their lives and work with their deepest values. At the organizational level, our publications promote progressive leadership and management practices, socially responsible approaches to business, and humane and effective organizations. At the societal level, our publications advance social and economic justice, shared prosperity, sustainable development, and new solutions to national and global issues.

We publish groundbreaking books focused on each of these levels. To further advance our commitment to positive change at the societal level, we have recently expanded our line of books in this area and are calling this expanded line "BK Currents."

A major theme of our publications is "Opening Up New Space." They challenge conventional thinking, introduce new points of view, and offer new alternatives for change. Their common quest is changing the underlying beliefs, mindsets, institutions, and structures that keep generating the same cycles of problems, no matter who our leaders are or what improvement programs we adopt.

We strive to practice what we preach—to operate our publishing company in line with the ideas in our books. At the core of our approach is stewardship, which we define as a deep sense of responsibility to administer the company for the benefit of all of our "stakeholder" groups: authors, customers, employees, investors, service providers, and the communities and environment around us. We seek to establish a partnering relationship with each stakeholder that is open, equitable, and collaborative.

We are gratified that thousands of readers, authors, and other friends of the company consider themselves to be part of the "BK Community." We hope that you, too, will join our community and connect with us through the ways described on our website at www.bkconnection.com.

Be Connected

Visit Our Website

Go to www.bkconnection.com to read exclusive previews and excerpts of new books, find detailed information on all Berrett-Koehler titles and authors, browse subject-area libraries of books, and get special discounts.

Subscribe to Our Free E-Newsletter

Be the first to hear about new publications, special discount offers, exclusive articles, news about bestsellers, and more! Get on the list for our free e-newsletter by going to www.bkconnection.com.

Participate in the Discussion

To see what others are saying about our books and post your own thoughts, check out our blogs at www.bkblogs.com.

Get Quantity Discounts

Berrett-Koehler books are available at quantity discounts for orders of ten or more copies. Please call us toll-free at (800) 929-2929 or email us at bkp.orders@aidcvt.com.

Host a Reading Group

For tips on how to form and carry on a book reading group in your workplace or community, see our website at www.bkconnection.com.

Join the BK Community

Thousands of readers of our books have become part of the "BK Community" by participating in events featuring our authors, reviewing draft manuscripts of forthcoming books, spreading the word about their favorite books, and supporting our publishing program in other ways. If you would like to join the BK Community, please contact us at bkcommunity@bkpub.com.